Machine Learning in Healthcare

Advances and Future Prospects

Machine Learning in Healthcare

Advances and Future Prospects

by

Rishabha Malviya, PhD

Niranjan Kaushik, PhD

Tamanna Rai, M.Pharm

M. P. Saraswathy, MBBS, MD, DNB

Rajendra Awasthi, PhD

AAP | APPLE ACADEMIC PRESS

First edition published 2026

Apple Academic Press Inc.
1265 Goldenrod Circle, NE,
Palm Bay, FL 32905 USA

760 Laurentian Drive, Unit 19,
Burlington, ON L7N 0A4, Canada

CRC Press
2385 NW Executive Center Drive,
Suite 320, Boca Raton FL 33431

4 Park Square, Milton Park,
Abingdon, Oxon, OX14 4RN, UK

Apple Academic Press exclusively co-publishes with CRC Press, an imprint of Taylor & Francis Group, LLC

Library and Archives Canada Cataloguing in Publication

CIP data on file with Canada Library and Archives

Library of Congress Cataloging-in-Publication Data

CIP data on file with US Library of Congress

ISBN: 978-1-77964-000-0 (hbk)
ISBN: 978-1-77964-001-7 (ebk)

About the Authors

Rishabha Malviya, PhD
Associate Professor, Department of Pharmacy, School of Medical and Allied Sciences, Galgotias University, Greater Noida, Uttar Pradesh, India

Rishabha Malviya, PhD, has been working in the Department of Pharmacy, School of Medical and Allied Sciences, Galgotias University, for the past eight years. He earned his Bachelor of Pharmacy from Uttar Pradesh Technical University and his Master of Pharmacy in Pharmaceutics from Gautam Buddha Technical University, Lucknow, Uttar Pradesh. His PhD in Pharmacy focused on novel formulation development techniques.

Dr. Malviya's areas of interest include formulation optimization, nanoformulation, targeted drug delivery, localized drug delivery, and the characterization of natural polymers as pharmaceutical excipients. He has authored over 150 research and review papers published in national and international journals, with a cumulative impact factor of 350. He holds 51 patents (12 granted, 38 published, and 1 filed).

Dr. Malviya has edited, co-edited, or authored 58 books with esteemed publishers, including Wiley, CRC Press/Taylor & Francis Group, Apple Academic Press/Taylor & Francis Group, River Publishers Denmark, Springer Nature, IOP Publishing, and OMICS Publishing Group. Additionally, he has authored 95 book chapters. He is recognized in the world's top 2% scientist list for the years 2020, 2021, 2023 and 2024, as compiled by Elsevier BV and Stanford University.

Niranjan Kaushik, PhD
Professor, Department of Pharmacy, Galgotias University School of Medicine and Allied Sciences, Greater Noida, India

Niranjan Kaushik, PhD, is currently a Professor in the Department of Pharmacy at Galgotias University's School of Medicine and Allied Sciences. With 15.5 years of research experience, he has edited two books and published more than 15 research and review papers in high-quality national and international journals. Additionally, he holds two granted patents, and four patents are published and undergoing assessment. Dr. Kaushik earned his doctorate in Pharmaceutical Chemistry from IFTM University, Moradabad, Uttar Pradesh. He completed his Master's degree in Pharmacy from Shri Gopichand College of Pharmacy, Baghpat, Uttar Pradesh, and his Bachelor's degree in Pharmacy from IFTM University, Baghpat.

Tamanna Rai, MPharm
School of Medical and Allied Sciences, Galgotias University, India

Tamanna Rai completed her BPharm from Ram-Eesh Institute of Vocational and Technical Education, affiliated with Dr. A.P.J. Abdul Kalam Technical University (AKTU), Lucknow, and her MPharm from Galgotias University, Greater Noida, India. She has participated in numerous national and international conferences and has published manuscripts in international journals.

M. P. Saraswathy, MBBS, MD, DNB
Associate Professor of Microbiology, ESIC Medical College and Hospital, Chennai, India

M. P. Saraswathy, MBBS, MD, DNB, is an Associate Professor of Microbiology at ESIC Medical College, Chennai. Dr. Saraswathy is an executive council member of the Indian Association of Medical Microbiologists (IAMM) Tamil Nadu chapter for the term 2023–2026. She also serves as the Nodal Officer for the ICTC-ART Centre at ESIC Medical College and Hospital, Chennai. She is a Clinical Microbiologist actively involved in diagnostic work and serving as an antimicrobial steward.

Dr. Saraswathy has published a book in collaboration with CBS Publishers, *SARP Review of Microbiology 2017*. She has several national and international publications to her credit and has served as a reviewer for resources such as the *Photon Image-Based PG Entrance Exam Book*, *Target High: Comprehensive Book for Nurses*, *International Journal of*

Microbiology Research, and *Journal of Evolution of Medical and Dental Sciences*, among others.

An enthusiastic teacher, Dr. Saraswathy has guided numerous Indian Council of Medical Research (ICMR) Short-Term Studentship (STS) and other student projects. She has also organized several Continuing Medical Education (CME) programs.

Rajendra Awasthi, PhD
Associate Professor of Pharmaceutics, School of Health Sciences, UPES University, Dehradun, India

Rajendra Awasthi, PhD, is currently an Associate Professor of Pharmaceutics at the School of Health Sciences, UPES University, Dehradun, India. Dr. Awasthi's research interests include cutting-edge advancements in nanotechnology-based drug delivery systems. He is a technically adept academician and pharmaceutical scientist with over 17 years of professional experience.

To support his expertise in formulation development, Dr. Awasthi has authored more than 160 publications, including 20 book chapters, with a total impact factor of nearly 300 and an H-index of 28. He holds one Indian design patent and two Australian innovation patents. Dr. Awasthi has successfully supervised two awarded PhD scholars and more than sixteen postgraduate students as the main supervisor. He is currently supervising six PhD scholars focused on nano-drug delivery systems.

Dr. Awasthi maintains active research collaborations with renowned researchers from prestigious universities, including the University of São Paulo in Brazil, the University of Technology Sydney in Australia, and the University of Bath in England. His biographical information has been featured in *Who's Who in the World*®. Additionally, he has been recognized in the top 2% of scientists globally, as published by Stanford University, California, for his achievements in pharmaceutical research.

Contents

About the Book

Machine Learning in Healthcare: Advances and Future Prospects is an in-depth exploration of the adoption of machine learning techniques in healthcare. The book comprises several chapters, each focusing on a distinct area of implementing machine learning to enhance disease diagnosis, treatment, and health management.

Chapter 1 introduces machine learning paradigms, including supervised, unsupervised, and reinforcement learning, emphasizing their critical role in healthcare decision-support systems. Key algorithms such as Naive Bayes, Decision Trees, and Convolutional Neural Networks are examined for their effectiveness in diagnosing diseases such as breast cancer, diabetes, and Alzheimer's disease.

Chapter 2 explores machine learning-based cancer detection and therapy, while Chapter 3 focuses on machine learning methods for the detection and treatment of cardiovascular diseases, highlighting the use of Random Forest, Naive Bayes, and Decision Trees algorithms.

Chapter 4 describes an architecture for monitoring thyroid patient health status using machine learning techniques, including data-gathering methods, proposed feature models, and classifier approaches. Chapter 5 examines machine learning applications in smart wearable devices, covering use cases such as sleep health, seizure detection, stress detection, hydration monitoring, diabetic monitoring, and arrhythmia detection.

For predicting and treating diabetes, Chapter 6 reviews machine learning algorithms such as K-Nearest Neighbors (KNN), Support Vector Machines (SVM), and Logistic Regression. Chapter 7 delves into the application of machine learning and deep learning approaches to mental health conditions, including depression, schizophrenia, anxiety, and bipolar disorder, using EEG characteristics, cognitive testing, and structural neuroimaging. Finally, Chapter 8 discusses machine learning algorithms for electronic health record (EHR) phenotyping, addressing challenges such as temporality, label absence, and multi-modality in EHR data.

Machine Learning in Healthcare: Advances and Future Prospects serves as a comprehensive resource for researchers, practitioners, and industry experts working at the intersection of machine learning and healthcare.

Foreword

The simultaneous emergence of machine learning and healthcare is reshaping the industry. The book *Machine Learning in Healthcare: Advances and Future Prospects* by Dr. Rishabha Malviya explores the possibilities of machine learning across various healthcare sectors. It aims to bridge the gap between data science and medical practice, focusing on disease detection, personalized therapy, and holistic patient care. The book's visionary curation underscores the transformative potential of machine intelligence in shaping the future of healthcare delivery.

The potential applications of machine learning algorithms in healthcare are vast, ranging from analyzing electronic health records to enabling predictive analytics for disease prevention and management. The book addresses ethical considerations, challenges, and the crucial collaboration between healthcare professionals and data scientists, emphasizing the need for responsible innovation in healthcare.

This book stands as a testament to the collective expertise and dedication of researchers in the field, serving as a beacon of knowledge for professionals navigating the evolving landscape of healthcare technology. It invites readers to immerse themselves in the profound implications of this technology and to contemplate the immense potential of machine learning in healthcare.

Dr. Malviya's leadership in compiling this invaluable resource is commendable. It is hoped that the insights within this book will inform and inspire future endeavors, catalyzing a new era in which machine learning becomes an indispensable ally in achieving healthier societies.

—Dr. Dhruv Galgotia
CEO, Galgotias University, Greater Noida, Uttar Pradesh

Preface

The integration of machine learning into healthcare has transformed technology for disease diagnosis, treatment, and management. This book explores the intricate relationship between data science and medical science, highlighting the significant impact of machine learning algorithms on various areas of healthcare. The convergence of traditional and deep learning paradigms offers a glimpse into a future where predictive analytics and decision support systems will revolutionize healthcare delivery, showcasing the promise of machine learning in the medical industry.

Each chapter exemplifies the vast possibilities enabled by computational technologies. These applications range from analyzing electronic health information to detecting and treating cancer, cardiovascular diseases, thyroid disorders, and diabetes.

Additionally, the exploration extends beyond conventional domains. Specific chapters focus on wearable devices and mental health management, illustrating how machine learning enhances mental health care and health monitoring.

While celebrating these achievements, this book also highlights the challenges that lie ahead. It acknowledges the complexity of data, the necessity of ethical considerations and interpretability, and the importance of collaboration between healthcare practitioners and data scientists to address these issues.

The book serves not only as a comprehensive guide for practitioners, scholars, and students but also as an invitation—an invitation to delve deeper, explore novel possibilities, and contribute to a future where machine learning becomes an invaluable partner in building healthier communities worldwide.

This book aspires to be a guiding light in this collaborative journey between technology and healthcare, inspiring creativity, fostering innovation, and catalyzing a paradigm shift in how we envision and practice healthcare.

Welcome to the convergence of machine learning and healthcare—a transformative union with the potential to reshape the future of human well-being.

CHAPTER 1

Machine Learning Algorithms in Disease Diagnosis and Management

ABSTRACT

Machine-learning algorithms have shown considerable promise in detecting many diseases due to their ability to analyze large data sets and establish conclusions. These algorithms use statistical models to learn from labeled examples and then apply that knowledge to new, unlabeled data. One major benefit of machine learning algorithms in disease diagnosis is their potential to increase accuracy and efficiency in detecting and diagnosing diseases. For example, they can analyze medical images, such as MRIs or CT scans, and identify abnormalities that may be difficult for human clinicians to detect. However, there are also challenges associated with using machine learning algorithms in healthcare. One challenge is the need for large amounts of high-quality data to train the algorithms. Another challenge is the potential for bias in the data, which can lead to inaccurate or unfair predictions. Despite these challenges, machine learning algorithms have the potential to revolutionize disease diagnosis and improve patient outcomes. They can help healthcare providers make faster and more accurate diagnoses, which can lead to earlier treatment and better outcomes for patients. They can also assist with personalized treatment plans by analyzing patient data and identifying the most effective treatments for everyone. Some potential applications of machine learning algorithms for disease diagnosis include predicting the likelihood of a patient developing a particular disease, identifying the best treatment plan for a patient, and predicting the efficacy of a particular treatment. Although there are obstacles that must be overcome, the benefits of

Machine Learning in Healthcare: Advances and Future Prospects. Rishabha Malviya, Niranjan Kaushik, Tamanna Rai, M. P. Saraswathy, and Rajendra Awasthi (Authors)

increased accuracy, efficiency, and personalized treatment plans make the investment in this technology worthwhile. Modern strategies for using machine learning in medical diagnostics consider an algorithm, disease categories, data types, applications, and evaluation methods. This chapter explores the use of machine learning to improve early disease detection and discusses remarkable discoveries and machine learning-based disease diagnosis trends and prospects.

1.1 INTRODUCTION

The term "machine learning," a subfield of artificial intelligence (AI), refers to a set of cutting-edge computational technologies that have found applications in a diverse range of fields, including the healthcare sector. There is a growing trend among individuals to utilize machine learning tools for disease diagnosis and forecasting associated healthcare costs. Numerous empirical investigations and real-world applications have demonstrated that machine learning-based disease diagnostics (MLBDD) exhibit significant promise as an economically viable and efficient approach for the identification of diseases [1]. The commonly employed conventional diagnostic techniques are characterized by their demanding nature in terms of labor, time consumption, and cost. Robotic systems possess the advantage of perpetual operation without experiencing fatigue, in contrast to machine learning-based systems that are not constrained by the user's proficiency level, as well as human doctors. Consequently, it is possible to design a therapeutic approach for managing diseases in regions characterized by a limited number of patients. Medical records, encompassing both visual representations such as MRI and X-ray images, as well as structured data in the form of tables containing information about patients' diseases, age, and gender, are utilized in the development of MLBDD systems [2]. Machine learning requires massive amounts of historical data to function properly [3].

Data scientists frequently employ specialized mathematical functions to accomplish complex objectives. Machine learning has the potential to enhance the efficiency of cancer cell detection in microscopic images. A deep learning (DL) study has demonstrated that the accuracy of MLBDD may exceed 90% [1]. Machine learning has been associated with the diagnosis and treatment of breast cancer, as well as its impact on kidney, liver, and heart diseases. The recent widespread adoption of predictive algorithms in the domain of disease detection serves as evidence of the potential advantages that this technology might offer in the healthcare sector. Some of the current challenges in the field of machine learning include addressing unbalanced data, interpreting machine learning models, and addressing ethical concerns in its application in medical domains. This chapter provides a comprehensive overview of machine learning and DL techniques and architectures that are utilized for the identification and classification of diseases. The aim is to enhance our understanding of the present trends, approaches, and limitations in the field of machine learning [4].

1.2 BACKGROUND

Machine learning, being a multidimensional subject that encompasses various disciplines such as statistics, mathematics, data management, and knowledge analytics, presents a formidable task in providing a single term [5]. Machine learning refers to the branch of AI that enables systems to acquire knowledge and improve performance by analyzing and interpreting real-world data. Figure 1.1 illustrates a range of subfields within the discipline of machine learning [6].

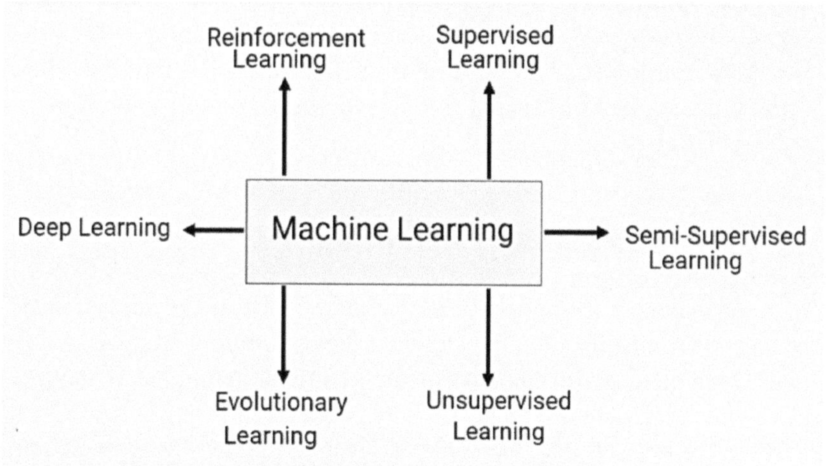

FIGURE 1.1 Types of machine learning techniques.

1.2.1 SUPERVISED MACHINE LEARNING

The utilization of data occurs throughout the training process, wherein an algorithm generates queries together with their corresponding answers, considering the available information. The utilization of supervised learning to accomplish classification problems has become a prevalent and widely accepted approach in various domains [7].

1.2.2 SEMI-SUPERVISED LEARNING

The technique involves the identification of unlabeled data that will yield the most advantages in enhancing the training process of a classifier. By using uncategorized data, it demonstrates a superior ability in classification. To ensure the effectiveness of this strategy, it is imperative to consider some underlying assumptions that have not been explicitly expressed.

1.2.3 UNSUPERVISED MACHINE LEARNING

During the initial stages of the unsupervised learning procedure, the learner lacks access to any form of labeled data. Unsupervised learning comprises a diverse array of approaches, such as hierarchical clustering, fuzzy clustering, K-means clustering, and association rule mining [8]. This classification is derived from the unlabeled training dataset. Structures can be constructed through the utilization of algorithms and representative data.

1.2.4 REINFORCEMENT LEARNING (RL)

In reinforcement learning (RL), computer software is utilized in this educational modality, providing learners the opportunity to engage with the interactive elements of the environment and facilitating the achievement of the intended outcome. The software is provided with reinforcement in the form of rewards and punishments for its progress toward overcoming the challenge [9].

1.2.5 EVOLUTIONARY LEARNING

The phenomenon of biological evolution may be conceptualized as a type of cognitive acquisition, as it enhances an organism's prospects for survival and procreation. The implementation of this model on a computer can be achieved by utilizing the concept of fitness to assess the accuracy of the response [10].

1.2.6 DEEP LEARNING (DL)

This subfield of machine learning is constructed on the fundamental principles of algorithms. The data manipulated by these learning algorithms is designed to mimic high-level abstraction to the greatest extent possible. The utilization of diverse linear and nonlinear transformations is a key aspect of the deep graph processing employed by the system.

The utilization of machine learning methods facilitated the analysis of medical database systems. In recent years, there has been a notable increase in the utilization of digital technology, resulting in reduced expenses and streamlined procedures for data gathering and storage. Machine learning has proposed several data analysis systems that exhibit enhanced efficiency. The utilization of data collection and processing devices, which is prevalent in modern healthcare institutions, enables the seamless exchange of patient information across extensive databases. The application of machine learning in medical data processing has proven advantageous for diagnostics. In modern healthcare facilities, there exist specialized sections dedicated to data management that are entrusted with the task of meticulously compiling precise diagnostic information inside patient records. Accurate input from diagnostic patient records is crucial for the optimal functioning of algorithms. The results of previous events can be determined in an automated manner. This classifier can be utilized by clinicians to expedite accurate diagnoses when encountering new patients. According to the cited source, individuals who are not experts, including students, can effectively identify issues with the assistance of these classifiers. Machine learning encompasses several applications such as voice identification, self-driving automobiles, web search, and generational perception. Due to its widespread presence in modern society, individuals may inadvertently employ machine learning techniques without conscious awareness. Machine learning is a field of study that investigates electronic health information to identify complex patterns and analyze different data sets. Pattern recognition plays a crucial role in the field of machine learning and technology (MLT) by facilitating assistance in both the prediction and planning phases of diagnosis and therapy [12]. Machine learning algorithms possess the capacity to effectively handle large volumes of data, integrate data from diverse sources, and incorporate prior knowledge into research activities.

1.3 THE HEALTHCARE INDUSTRY REQUIRES A DECISION SUPPORT SYSTEM

Medical errors are responsible for a considerable number of fatalities annually in the United States, with a substantial number of individuals also sustaining injuries because of these errors. The health information technology framework proposes a range of techniques, including incorporating consumer comprehension into the physician and organization selection process, fostering collaboration, and facilitating the adoption of IT [13].

1.3.1 DECISION SUPPORT

The optimal performance of machine learning-based medical security systems is achieved through a collaborative partnership between healthcare professionals and computational technology. The goal of this process is to attain the utmost level of productivity. Simultaneous monitoring of heart rates for all patients and diagnosing all disorders is unattainable for both machines and physicians. Both the machine and the physician are actively searching for a common denominator, although neither has been successful in locating it. Upon completion of data processing by the machine, the outcomes of the analysis will be presented to the physician for examination.

1.3.2 DECISION SUPPORT SYSTEM IN HEALTHCARE

The implementation of a decision support tool will provide more visibility on financial information to the staff of a clinic regarding patient invoicing, payments, and related expenditures. This strategy, in addition to supporting the patient in preserving insurance coverage, offers several possibilities for repayment. It provides numerous decision support system modules for use in the healthcare industry [14]. Examples of how decision support systems aid in the study of diseases include the compilation of medical experts' opinions on various health-related topics and the disclosure of medical records for patients. This web-based system is linked to electronic health data, which enables it to serve as a medication scheduler in addition to assisting in patient diagnosis.

1.4 MACHINE LEARNING ALGORITHMS

The most popular and widely used machine learning algorithm-based clinical diagnosis techniques are described in subsections.

1.4.1 NAIVE BAYES (NB)

Naive bayes (NB) is a Bayesian probabilistic classifier that is a relatively new method. Upon obtaining a single record or fragment of data, the computer software will proceed to engage in a process that evaluates the probability that such a record or component will be assigned to each respective category. Based on the calculations, it may be inferred that the outcome with the greatest potential score is the most probable. The NB classifier does not offer predictions; instead, it generates probabilistic forecasts [15].

1.4.2 SUPPORT VECTOR MACHINE (SVM)

Support vector machines (SVMs) are frequently employed in machine learning for various purposes, such as classification and regression tasks. Vapnik invented SVM in the second half of the 20th century [16]. In addition to its applications in the medical field, SVM has been utilized in many domains, such as speech recognition, facial expression identification, protein folding, identification of distant homologies, and text categorization. The performance of a supervised machine learning algorithm is likely to be suboptimal when applied to unlabeled data. SVM employs a hyperplane to identify patterns and groupings within unlabeled data, thereby facilitating the classification process. Currently, it is not possible to do nonlinear partitioning of the results obtained from an SVM analysis. Before utilizing an SVM implementation for data analysis, it is imperative to meticulously choose an appropriate kernel and a corresponding set of parameters [15].

1.4.3 K-NEAREST NEIGHBOR (KNN)

In 1951, Evelyn Fix and Joseph Hodges introduced a nonparametric classification method known as k-nearest neighbor (KNN). The KNN algorithm is capable of performing both classification and regression tasks. The KNN algorithm is utilized to classify class membership. The voting method classifies the item. Techniques grounded in Euclidean distance can be employed to ascertain the gap between two datasets. The anticipated value in a regression analysis is determined by calculating the average of the KNN values [17].

1.4.4 ADABOOST

Yoav Freund and Robert Schapiro constructed the AdaBoost algorithm. AdaBoost is a classification algorithm that combines the most effective features from multiple classifiers to create a consolidated and more precise model. AdaBoost prioritizes samples that are more challenging to categorize and less prioritizes those that are simpler. The program can be utilized for both categorization and statistical analysis with the implementation of regression [18].

1.4.5 DECISION TREE (DT)

The decision tree (DT) algorithm is based on the principle of "divide and conquer." In DT models, the representation of "classification trees" entails the depiction of categories as terminal nodes, commonly referred to as leaves, while the elements that contributed to their determination are represented as branches. Regression trees, on the other hand, are a type of continuous variable that can be used with DT. The C4.5 and

EC4.5 algorithms have gained significant recognition and are commonly employed in the field of DT algorithms [19–23].

1.4.6 FUZZY LOGIC

The concept of fuzzy sets served as the foundation for the development of fuzzy logic. These numbers are hypothetical and fall between the ranges of zero and one. This methodology is commonly employed in the field of engineering [24].

1.4.7 CLASSIFICATION AND REGRESSION TREE

Classification and regression trees (CART) use categorical or continuous objective variables. The prediction of values in the tree can be done by considering these factors [25].

1.4.8 LOGISTIC REGRESSION (LR)

The logistic regression (LR) technique is employed to address classification problems. The LR model is based on the concept of probability, where the predicted values are confined within the range of 0 to 1. LR-based machine learning has a wide range of applications, including the detection of spam emails, the identification of fraudulent online transactions, and the diagnosis of malignant tumors. The cost function utilized in LR is represented by a sigmoid function. The sigmoid function is a mathematical transformation that maps all real values within the range of 0 and 1 [26].

1.4.9 CONVOLUTIONAL NEURAL NETWORK (CNN)

The utilization of convolutional neural networks (CNNs) is increasingly prevalent in the field of image processing. The CNN exhibits a wide range of applications, including biological image detection and recognition, face recognition, text analysis, and organ localization [27]. Since its inception in 1989, CNN has witnessed the development of a novel variation that has demonstrated remarkable efficacy in the field of disease diagnosis. A conventional CNN architecture typically comprises three distinct layers, namely the input layer, the hidden layer, and the output layer. The hidden layers within a feedforward network serve as intermediary layers. The number of hidden layers present within a given structure can exhibit variability. In the process of hidden layer convolution, the dot products between the convolution kernel and the input matrix are preserved. Following the convolutional layers, the output of the previous layer is used as the input for the subsequent layer. Upon successfully finishing the first stage, additional layers are shown for completion. Two examples that might be mentioned in this context are the pooling layer and a fully connected layer [28]. A variety of CNN models have been reported. Figure 1.2 provides a visual representation of CNN models, which have gained significant popularity and widespread acceptance within the research community.

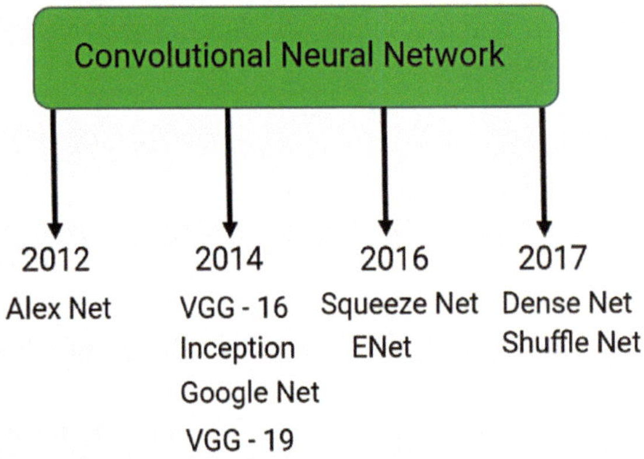

FIGURE 1.2 Convolutional neural network models and their development timeline.

1.5 DEEP LEARNING (DL)

Deep learning (DL) uses hierarchical structures to acquire knowledge from numerical significance, occurrences, and categorization. CNNs are widely utilized in contemporary DL structures, showcasing their versatility in being seamlessly included in generative models, deep neural networks (DNNs), and Boltzmann machines. Three major categories of DL methodologies are supervised, semi-supervised, and unsupervised approaches. DNNs, RLs, and recurrent neural networks (RNNs) are widely used DL structures [28]. To facilitate the advancement of progressive layers in DL, each subsequent layer needs to ascertain the means to transform its input data into the specific format mandated by the layer positioned above it. In the context of image recognition applications, it is common for the initial layer of a neural network to be responsible for detecting edges inside a given pixel matrix. The subsequent layer will construct and encode the ocular and nasal features, and assuming successful execution, the next layer will identify the facial structure by incorporating data from both preceding levels [29]. The potential for DL to improve healthcare is significant. DL has been widely employed in the fields of radiology and pathology to facilitate disease diagnosis [30]. Further investigation is required to examine the practical applications of DL in human research. This includes the collection of molecular state data and the monitoring of illness progression or treatment sensitivity [31].

1.6 PERFORMANCE EVALUATIONS

This section explores some popular ways to measure performance. Disease diagnosis often makes use of metrics including precision, recall, accuracy, and the F1-score. For example, correct lung cancer diagnoses are true positive (T_P) or true negative (T_N), while faulty diagnoses are false positive (F_P) or false negative (F_N). Some of the most common measurements are outlined in subsections [32].

1.6.1 ACCURACY

Accuracy represents the proportion of cases in which the identification is correct. Accuracy is determined using the following formula:

$$\text{Accuracy} = \frac{T_p + T_N}{T_p + T_N + F_p + F_N}$$

1.6.2 PRECISION

Precision is quantified as the ratio of accurately anticipated events to the total number of successfully predicted events.

$$\text{Precision} = \frac{T_p}{T_p + F_p}$$

1.6.3 RECALL

Recall measures how many relevant results the algorithm gets right.

$$\text{Recall} = \frac{T_p}{T_{n+F_p}}$$

1.6.4 F-MEASURE

The F-score, also known as the F-measure, is calculated as the harmonic mean of the accuracy and recall scores. Assuming perfect accuracy and recall, an F score of 1 is the best that can be obtained.

$$\text{F-Measure} = 2 \times \frac{\text{Precision} \times \text{Recall}}{\text{Precision} + \text{Recall}}$$

1.6.7 AREA UNDER CURVE

The calculation of the area under the curve provides insight into the performance of models across different situations. The area under the curve can be computed using following formula:

$$\text{Area under curve} = \frac{\sum Ri\left(I_p\right) - I_p((I_p + 1)/2)}{I_p + I_n}$$

1.6.8 SPECIFICITY

Specificity identifies how many true negatives (T_N) are appropriately identified. Specificity is calculated as follows:

$$\text{Specificity} = \frac{T_N}{T_N + F_P}$$

1.7 PREDICTION OF DISEASE OUTCOMES USING MACHINE LEARNING

1.7.1 MACHINE LEARNING IN THE DETECTION OF BREAST CANCER

Breast cancer is a common form of cancer among women and ranks as the second leading cause of death in the United States and Asian nations. Several machine learning algorithms have demonstrated the ability to accurately predict a breast cancer diagnosis. The University of California, Irvine (UCI) ML library provided the Wisconsin data. In their study, Williams et al. employed a J48 DT using a NB model to assess the risk factors associated with breast cancer across the United States. WEKA is used to conduct experiments. According to a study, the J48 algorithm has been identified as the breast cancer prediction algorithm with the highest accuracy rate of 94.2%, surpassing the NB algorithm, which had an accuracy rate of 82.6% [33]. To detect breast cancer at an early stage, Senturk et al. employed machine learning algorithms including NB, SVM, KNN, and DT. SVM had an accuracy of 96.4%, whereas KNN had 95.15% [34]. Amaryeen et al. used DTs and data mining trends to predict breast cancer. DTs exhibit a high accuracy rate of 94% [35].

1.7.2 MACHINE LEARNING IN THE DETECTION OF DIABETES

Iyer et al. predicted diabetes using DTs and NB. Insulin deficiencies or improper use can result in the progression of diabetes. The Pima Indian Diabetes Data Set was utilized in the research conducted in this particular field. A range of experiments was conducted on the data mining tool WEKA to ascertain its reliability and efficacy. It has been observed that the utilization of a 70:30 percentage split yields superior performance compared to cross-validation when applied to this specific data set. The accuracy of the J48 classifier was found to be 74.8698% when evaluated using cross-validation and 76.9565% when utilizing a percentage split.

The accuracy of the NB classifier demonstrates an increase to 79.5652% when incorporating the use of the percentage split technique. The degrees of accuracy achieved by algorithms are typically measured and presented in a split test, denoted as a percentage (%). This metric provides insights into the maximal performance capabilities of the algorithms under evaluation [36].

Sen and Dash address the use of meta-learning algorithms in the diagnosis of diabetes. The dataset pertaining to diabetes among Pima Indians was collected from the UCI Machine Learning Laboratory. The analysis is conducted using the WEKA software. The prediction of a patient's diabetes state is conducted by the utilization of several machine learning algorithms, including CART, AdaBoost, Logiboost, and grading learning methods. Experimental findings are compared using both accurate and inaccurate classifications. The classification accuracy achieved by the CART algorithm is 78.64% [37]. The AdaBoost algorithm achieved a degree of accuracy of 77.864%. The Logiboost algorithm yielded an accuracy rate of 77.479%. The grading method demonstrates a considerable level of accuracy, with a categorization rate of 66.406%. The classification error rate of the CART algorithm is observed to be 21.354%, indicating a comparatively reduced misclassification rate when compared to alternative approaches. CART has the potential to achieve a maximum accuracy of 78.64%.

Sarwar and Sharma proposed the application of the NB algorithm as a predictive model for type 2 diabetes. Diabetes has the potential to present itself in one of three distinct forms. Type 1 diabetes is the most commonly observed form, followed by type 2 diabetes, and lastly, gestational diabetes. Type 2 diabetes is characterized by an increase in insulin resistance, leading to its development. The dataset provided comprised 415 samples, which were carefully selected to encompass a diverse variety of demographic characteristics within the Indian population. The development of the model involved the use of MATLAB and SQL Server. The application of NB yielded a forecast accuracy rate of 95% [38].

The integration of genetic algorithms and fuzzy logic has been used to detect and classify diabetes. This technique enhances the accuracy of classification and facilitates the selection of an optimal set of features. The datasets utilized in the trials were sourced from the UCI Machine Learning Laboratory and comprise 769 instances, each characterized by 8 distinct attributes. MATLAB is used to develop applications. A genetic algorithm selects only the top three features or attributes. A fuzzy logic classifier using these three features yields 87% accuracy. The revised price tag was approximately 50% of the original value.

A naïve Bayes-based technique can be used to diagnose diabetes more accurately. In 2012, the NB algorithm demonstrated the highest level of accuracy, reaching 95%. The results indicate that the system exhibits a high level of precision in its predictions, with a minimal margin of error. Furthermore, the methodology employed is essential for the detection of diabetes in individuals. However, the accuracy of NB was considered suboptimal in 2015, with a precision percentage of 79.5652%.

1.7.3 MACHINE LEARNING IN DETECTION OF KIDNEY DISEASE

Patients with renal disease have compromised kidney function, and if the problem is not treated promptly, it may result in kidney failure. Based on data provided by the National Kidney Foundation, it is estimated that around 10% of the global population is afflicted with chronic kidney disease, resulting in a significant number of fatalities each year. The utilization of machine learning and DL techniques for the detection of kidney disease holds the potential to assist countries facing challenges in managing the diagnosis of renal illnesses [40]. In their study, Charleonnan et al. [41] employed publicly available datasets to assess the performance of various classification algorithms. The KNN classifiers achieved an accuracy rate of 98.1%, while the SVM classifiers achieved an accuracy rate of 98.3%. LR classifiers demonstrated an accuracy rate of 96.55%, and DT classifiers achieved an accuracy rate of 94.8% [41]. Aljaaf et al. [42] explored similar research using various machine learning algorithms, including RPART, SVM, LOGR, and MLP, which were applied to the chronic kidney disease dataset. Among these algorithms, MLP demonstrated the highest level of effectiveness, with an accuracy rate of 98.1% [42]. Ma et al. employed a diverse range of datasets sourced from several channels to conduct the diagnosis of chronic renal disease [43, 44]. The researchers were able to attain a high level of accuracy, ranging from 87% to 99%, by employing their suggested HMANN model, which consists of heterogeneous modified artificial neural networks (ANNs). Table 1.1 summarizes the studies highlighting the application of machine learning in the diagnosis of kidney diseases [45–47].

TABLE 1.1 The Study Highlighted Research on Machine Learning-Based Kidney Disease Diagnostics

Algorithm	Dataset	Contributions	Data Type	Performance Evaluation
CNN-SVM	Privately owned dataset	Chronic kidney disease	Tabular	Sensitivity – 97%, Specificity – 97.8%
CNN	Privately owned data	Detection and localization of kidneys in patients with autosomal dominant polycystic	Image	Accuracy – 94%
LR, Feedforward NN, and Wide DL	Chronic kidney disease dataset	Classification of chronic kidney disease	Tabular	Precision – 97%, Recall – 99%, AUC – 99%
ANN and Kernel KMC	Data from patient ultrasounds	Kidney disease detection and segmentation	Image	Accuracy – 99.61%
NB, DT, and RF	Chronic kidney disease dataset	Analysis of chronic kidney disease	Tabular	Accuracy – 100% (RF)

1.7.4 MACHINE LEARNING IN DETECTION OF LIVER DISEASE

Vijayarani and Dhayanand have successfully employed the SVM and NB classification methods to obtain precise prognostications pertaining to liver disease [48]. The Indian liver patient dataset was used from the UCI database collection. The collection consists of 560 cases and encompasses 10 distinct attributes. The criteria of accuracy and speed of execution are employed for the purpose of comparison. The NB classifier achieved an accuracy rate of 61.28% with a processing time of 1670.00 milliseconds. The SVM algorithm resulted in an accuracy rate of 79.66% within a time frame of 3210.00 milliseconds. Actualization was accomplished with the help of MATLAB. When conducting a comparative analysis between NB and SVM, it was observed that SVM exhibited superior accuracy in predicting liver disease. The computational efficiency of NB surpasses that of SVM.

Rajeswari and Reena used data mining methods including NB, K-star, and FT trees to study liver disease. From that location, 345 instances and seven attributes of the UCI dataset were extracted. The WEKA software is used for the execution of 10 distinct cross-validation tests. The NB algorithm demonstrates a high level of accuracy, with a correctness rate of 96.52% within an instantaneous time frame. The FT Tree algorithm has the capability to achieve an accuracy rate of 97.10% within a time frame of less than one second. The K-star algorithm demonstrates a sorting capability for incoming instances with an accuracy rate of 83.47%, achieving this result in no time. In comparison to alternative data mining techniques, FT trees exhibit superior classification accuracy when applied to the liver disease dataset [49]. In comparison to alternative algorithms, the FT Tree algorithm demonstrates superior performance in the diagnosis of liver illness. The application of the FT tree approach to the liver disease dataset leads to a reduction in the time required to generate the model. Based on its inherent attributes, it exhibits a higher level of efficiency. This method demonstrates a comprehensive attribute categorization approach, achieving a notable level of accuracy at 97.10%. The algorithm plays a pivotal role in determining the high classification accuracy of the dataset based on the results.

1.7.5 MACHINE LEARNING IN DETECTION OF HEART DISEASE

Machine learning is employed in both research and clinical settings for diagnosing heart diseases. Ansari et al. [50] proposed a neurofuzzy integrated-systems-based automated coronary heart disease diagnosis system that reaches around 89% accuracy [51, 52]. The authors fail to address the effectiveness of their methodology in various additional scenarios, including multiclass classification, handling big datasets, and dealing with imbalanced class distributions. This omission was a significant limitation of their study. However, the discussion surrounding the validity of the model was hardly addressed, despite the prevalent recommendation to engage in such discussions within the field of modern medicine. The ability to comprehend the procedure by others outside the medical domain confers a significant advantage, therefore rendering it of utmost importance.

Deep CNNs developed by Rubin et al. are used to detect abnormal heart sounds. During the training process on this dataset, the loss function is modified to enhance its sensitivity and specificity. This proposed model was submitted as part of the 2016 PhysioNet computing competition. Overall, it was ranked second due to its specificity of 0.95 and sensitivity of 0.73 [52]. Moreover, there has been an increasing interest in the utilization of DL-based algorithms to diagnose cardiac diseases. An example of a technique in the field of cardiotocographic fetal health detection is the DL-based approach proposed by Miao and Miao. This technique utilizes a multiclass morphologic pattern for accurate classification. The DL-based model proposed by Miao and Miao is employed for identifying and classifying the physiological features of expectant females encountering challenges. The initial computational study yielded an accuracy rate of 88.02%, a precision rate of 85.01%, and an F-score of 0.85 [53]. The authors of the study utilized various dropout methods to mitigate the issue of overfitting, resulting in a decrease in training speed but eventually leading to an enhancement in accuracy. Despite the widespread use of machine learning applications for this specific objective, there exists a dearth of scholarly investigations that have specifically tackled the obstacles associated with multiclass classification in the presence of imbalanced data. It is expected that the model will not be able to explain its final forecast. Table 1.2 provides a comprehensive overview of the scholarly literature pertaining to the use of machine learning and DL techniques in the domain of cardiac diagnostics [32, 54–60].

TABLE 1.2 Referenced Literature That Considered Machine-Learning-Based Heart Disease Diagnosis

Contributions	Dataset	Data Type	Algorithms	Performance Evaluation
Predicting heart disease	Cleveland dataset	Tabular	RF, CNN	F1-score – 80%; CNN: accuracy – 78.8%, precision – 80%
Heart disease classification	Cleveland dataset	Tabular	SVM, back-propagation	Accuracy – 73%
CHD prognosis	Cleveland dataset	Tabular	Gaussian NB, Bernoulli NB, and RF	BNN – 85.08%, RF – 85%
Heart disease classification	Cleveland dataset	Tabular	NN, LR	Accuracy: NN – 85.6%, LR – 92%
Intelligent scoring for 72-hour cardiac arrest prediction	Privately owned	Tabular	CNN	Sensitivity – 62%, positive predictive value – 10%, negative predictive value – 98%
New heartbeat recognition method	MIT-BIH	Tabular	SVM	Accuracy – 97% (noise-free ECGs)

1.7.6 MACHINE LEARNING IN THE DETECTION OF ALZHEIMER'S DISEASE

Alzheimer's disease is recognized as the primary cause of dementia [61]. It is a progressive neurodegenerative disorder that disproportionately affects a significant proportion of the older population, ranging from 60% to 70%. Alzheimer's disease has been shown to manifest in various cognitive and behavioral symptoms, including but not limited to language impairment, confusion, emotional instability, and atypical conduct. The decline in physiological functioning exhibits a slow progression, with the median duration of survival after diagnosis varying between 3 and 9 years. Early detection could potentially enhance the likelihood of survival, as it allows medical professionals to implement preventive interventions and initiate suitable therapeutic measures. Over time, the use of machine learning and DL techniques for diagnosing Alzheimer's disease has demonstrated encouraging results. Neelaveni and Devasana utilized SVM and DT as classification algorithms to differentiate individuals with Alzheimer's disease from control subjects. The SVM and DT achieved accuracy rates of 83% and 85%, respectively [62]. In their study, Collij, and coworkers used SVMs to predict the occurrence of Alzheimer's disease and moderate cognitive impairment (MCI) in individuals. Numerous endeavors have been made to enhance the diagnosis of Alzheimer's disease through the utilization of machine learning techniques, with a wide array of algorithms being explored and evaluated [63]. The study conducted by Vidushi and Shrivastava [64] revealed that LR, SVM, DT, and ensembles of random forests (RFs) exhibited accuracy rates ranging from 78.95% to 84.21%, 81.58% to 84.21%, and 84.21% to 84.21%, respectively. CNN has been employed in numerous studies focused on the identification of Alzheimer's patients because of its superior performance compared to alternative image processing techniques [64]. The CNN model suggested by Ahmed et al. aims to detect and classify the early stages of Alzheimer's disease. The accuracy of the model, which was trained on a dataset consisting of 6628 MRI images, was reported to be 99% [65]. The deep feature-based models proposed by Nawaz et al. demonstrated a high accuracy rate of 99.12%. This finding further supports the effectiveness of a CNN-based technique in the diagnosis of Alzheimer's disease [66]. Table 1.3 presents a comprehensive overview of the several machine learning and DL methodologies that are presently employed in the diagnosis of Alzheimer's disease.

TABLE 1.3 Recognized Literature on Machine Learning-Based Alzheimer's Disease Diagnosis

Contributions	Algorithms	Data Type	Dataset	Performance Evaluation
Understanding the development of moderate cognitive impairment to Alzheimer's disease	LR, ARN, DT	Image	1913 privately owned cases	Sensitivity – (82.11 ± 0.36%), Positive predictive value – (75.26 ± 0.86%)
Automatic classification of Alzheimer's	DNN + RF	Tabular	–	Accuracy – 67%
Automatic diagnosis of Alzheimer's disease and mild cognitive impairment	CNN + SVM	Image	F-FDG PET dataset: PET	Accuracy – 74–90%

1.8 MEDICAL USES FOR MACHINE LEARNING

Machine learning algorithms can identify small yet crucial patterns among extensive and diverse datasets. This technique is expected to provide significant assistance in therapeutic applications, particularly those that rely on high-throughput measurements of genomes and proteomics. It is very useful in medicine, particularly in identifying and diagnosing conditions in certain individuals. Machine learning algorithms have the potential to enhance the decision-making capabilities of medical professionals and offer insights for optimizing the functionality of the healthcare system [6]. The utilization of this methodology by healthcare industry managers is aimed at approximating the duration of patients' waiting periods within the emergency room. These models incorporate several factors, such as patient data, pain ratings, emergency room charts, and the layout of the waiting area, to make estimations regarding wait times. Healthcare facilities utilize prognostic models to strategically plan for inpatient hospital stays. Consequently, patients can potentially enjoy advantages from machine learning applications through cost reduction, enhanced precision, or the widespread availability of time-limited opportunities.

1.8.1 PROBLEMS ASSOCIATED WITH ALGORITHMS

The majority of machine learning models exhibited remarkable performance when trained on labeled data, surpassing their unsupervised counterparts. The efficacy of such algorithms significantly decreased when confronted with unannotated data. The efficacy of widely recognized techniques such as K-means clustering, SVMs, and kernel-based nearest neighbors (KNNs) experienced a decrease in performance when employed on data with many dimensions. CNNs present challenges due to their opaque nature. A significant limitation of this approach lies in the lack of transparency regarding the mechanism by which the model adjusts its internal parameters, such as its learning rate and weights. The widespread utilization of algorithm-based models in healthcare necessitates the establishment of justifications for their implementation.

1.9 CONCLUSION

Machine learning can be utilized as a powerful tool by individuals engaged in the field of medicine, including practitioners, scholars, and students. It appears that there is a constant influx of fresh advancements in the field of machine learning on a daily basis. With the advent of each technological breakthrough, a novel machine learning application is arising, exhibiting the capacity to address tangible challenges within the healthcare sector. By employing artificial intelligence (AI)-based solutions and machine learning models, multinational corporations can deliver enhanced healthcare services to their clientele. This technology facilitates the efforts of organizations and pharmaceutical manufacturers in expediting and streamlining the process of developing treatments for severe diseases. Machine learning models have the capability to identify individuals at an elevated risk of chronic diseases, such as heart disease and renal disease, by employing a range of known algorithms specifically designed for this purpose.

KEYWORDS

- **convolutional neural networks**
- **decision tree algorithm**
- **deep learning**
- **fuzzy logic**
- **machine learning**
- **supervised learning**
- **support vector machine**

REFERENCES

1. Ahsan, M. M., & Siddique, Z. (2022). Machine learning-based heart disease diagnosis: A systematic literature review. *Artificial Intelligence in Medicine*, *129*, 102289.
2. Ponta, L., Puliga, G., Oneto, L., & Manzini, R. (2020). Identifying the determinants of innovation capability with machine learning and patents. *IEEE Transactions on Engineering Management*, *69*(5), 2144–2154.
3. Gill, L., Zarbo, A., Isedeh, P., Jacobsen, G., Lim, H. W., & Hamzavi, I. (2016). Comorbid autoimmune diseases in patients with vitiligo: A cross-sectional study. *Journal of the American Academy of Dermatology*, *74*(2), 295–302.
4. Ahsan, M. M., Gupta, K. D., Islam, M. M., Sen, S., Rahman, M. L., & Shakhawat Hossain, M. (2020). Covid-19 symptoms detection based on nasnetmobile with explainable AI using various imaging modalities. *Machine Learning and Knowledge Extraction*, *2*(4), 490–504.
5. Naim, A., Lahav, O., Sodre Jr, L., & Storrie-Lombardi, M. C. (1995). Automated morphological classification of APM galaxies by supervised artificial neural networks. *Monthly Notices of the Royal Astronomical Society*, *275*(3), 567–590.
6. Shailaja, K., Seetharamulu, B., & Jabbar, M. A. (2018). Machine learning in healthcare: A review. In *2018 Second International Conference on Electronics, Communication and Aerospace Technology (ICECA)* (pp. 910–914). IEEE.
7. Allix, N. M. (2003). Epistemology and knowledge management concepts and practices. *Journal of Knowledge Management Practice*, *4*(1), 1–24.
8. Prakitsuwan, P., & Moschis, G. P. (2021). Well-being in later life: A life course perspective. *Journal of Services Marketing*, *35*(1), 131–143.
9. Derman, C. (1970). *Finite state Markovian decision processes*. Academic Press, Inc.
10. Pavithra, D., & Jayanthi, A. N. (2018). A study on machine learning algorithm in medical diagnosis. *International Journal of Advanced Research in Computer Science*, *9*(4).
11. Brown, J. S., & Duguid, P. (2001). Structure and spontaneity: Knowledge and organization. In *Managing Industrial Knowledge: Creation, Transfer and Utilization* (pp. 44–67). SAGE Publications.
12. Rambhajani, M., Deepanker, W., & Pathak, N. (2015). A survey on implementation of machine learning techniques for dermatology diseases classification. *International Journal of Advances in Engineering & Technology*, *8*(2), 194.
13. Robinson, V. M. (2002). Organizational learning, organizational problem solving and models of mind. In *Second International Handbook of Educational Leadership and Administration* (pp. 775–812). Springer.
14. Rajalakshmi, K., Mohan, S. C., & Babu, S. D. (2011). Decision support system in healthcare industry. *International Journal of Computer Applications*, *26*(9), 42–44.
15. Akbar, S., Akram, M. U., Sharif, M., Tariq, A., & Khan, S. A. (2018). Decision support system for detection of hypertensive retinopathy using arteriovenous ratio. *Artificial Intelligence in Medicine*, *90*, 15–24.
16. Drucker, H., Wu, D., & Vapnik, V. N. (1999). Support vector machines for spam categorization. *IEEE Transactions on Neural Networks*, *10*(5), 1048–1054.

17. Ballerini, L., Fisher, R. B., Aldridge, B., & Rees, J. (2013). A color and texture based hierarchical K-NN approach to the classification of non-melanoma skin lesions. In *Color Medical Image Analysis* (pp. 63–86). Springer.
18. Schapire, R. E. (2013). Explaining AdaBoost. In *Empirical Inference: Festschrift in Honor of Vladimir N. Vapnik* (pp. 37–52). Springer.
19. Charbuty, B., & Abdulazeez, A. (2021). Classification based on decision tree algorithm for machine learning. *Journal of Applied Science and Technology Trends, 2*(01), 20–28.
20. Walse, R. S., Kurundkar, G. D., Khamitkar, S. D., Muley, A. A., Bhalchandra, P. U., & Lokhande, S. N. (2021). Effective use of naïve Bayes, decision tree, and random forest techniques for analysis of chronic kidney disease. In *Information and Communication Technology for Intelligent Systems: Proceedings of ICTIS 2020, Volume 1* (pp. 237–245). Springer Singapore.
21. Rajendran, K., Jayabalan, M., & Thiruchelvam, V. (2020). Predicting breast cancer via supervised machine learning methods on class imbalanced data. *International Journal of Advanced Computer Science and Applications, 11*(8).
22. Song, Y. Y., & Ying, L. U. (2015). Decision tree methods: Applications for classification and prediction. *Shanghai Archives of Psychiatry, 27*(2), 130.
23. Nurrohman, A., Abdullah, S., & Murfi, H. (2020). Parkinson's disease subtype classification: Application of decision tree, logistic regression and logit leaf model. In *AIP Conference Proceedings* (Vol. 2242, No. 1, p. 030015). AIP Publishing LLC.
24. Zimmermann, H. J. (2011). *Fuzzy Set Theory—and Its Applications*. Springer Science & Business Media.
25. Amato, F., López, A., Peña-Méndez, E. M., Vaňhara, P., Hampl, A., & Havel, J. (2013). Artificial neural networks in medical diagnosis. *Journal of Applied Biomedicine, 11*(2), 47–58.
26. Luu, B. C., Wright, A. L., Haeberle, H. S., Karnuta, J. M., Schickendantz, M. S., Makhni, E. C., Nwachukwu, B. U., Williams III, R. J., & Ramkumar, P. N. (2020). Machine learning outperforms logistic regression analysis to predict next-season NHL player injury: An analysis of 2322 players from 2007 to 2017. *Orthopaedic Journal of Sports Medicine, 8*(9), 2325967120953404.
27. Yap, M. H., Pons, G., Marti, J., Ganau, S., Sentis, M., Zwiggelaar, R., Davison, A. K., & Marti, R. (2017). Automated breast ultrasound lesions detection using convolutional neural networks. *IEEE Journal of Biomedical and Health Informatics, 22*(4), 1218–1226.
28. Goodfellow, I., Bengio, Y., & Courville, A. (2016). *Deep Learning*. MIT Press.
29. Faust, O., Hagiwara, Y., Hong, T. J., Lih, O. S., & Acharya, U. R. (2018). Deep learning for healthcare applications based on physiological signals: A review. *Computer Methods and Programs in Biomedicine, 161*, 1–3.
30. Hayashi, Y. (2019). The right direction needed to develop white-box deep learning in radiology, pathology, and ophthalmology: A short review. *Frontiers in Robotics and AI, 6*, 24.
31. Akkus, Z., Galimzianova, A., Hoogi, A., Rubin, D. L., & Erickson, B. J. (2017). Deep learning for brain MRI segmentation: State of the art and future directions. *Journal of Digital Imaging, 30*, 449–459.

32. Acharya, U. R., Oh, S. L., Hagiwara, Y., Tan, J. H., Adam, M., Gertych, A., & Tan, R. S. (2017). A deep convolutional neural network model to classify heartbeats. *Computers in Biology and Medicine*, *89*, 389–396.
33. Williams, K., Idowu, P. A., Balogun, J. A., & Oluwaranti, A. I. (2015). Breast cancer risk prediction using data mining classification techniques. *Transactions on Networks and Communications*, *3*(2), 01.
34. Turkyilmaz, Y., Senturk, A., & Bayrakdar, M. E. (2023). Employing machine learning based malicious signal detection for cognitive radio networks. *Concurrency and Computation: Practice and Experience*, *35*(2), e7457.
35. Al-Amaryeen, M. H., & Al-Majali, H. D. (2023). Disturbances identification by using machine learning algorithms. In *2023 IEEE Jordan International Joint Conference on Electrical Engineering and Information Technology (JEEIT)* (pp. 32–37). IEEE.
36. Iyer, A., Jeyalatha, S., & Sumbaly, R. (2015). Diagnosis of diabetes using classification mining techniques. *arXiv* preprint arXiv:1502.03774.
37. Sen, S. K., Pani, S. K., Ojha, A. C., & Dash, S. (2015). Meta-learning in data classification: An analysis. *IUP Journal of Computer Sciences*, *9*(1), 14.
38. Sarwar, A., & Sharma, V. (2012). Intelligent Naïve Bayes approach to diagnose diabetes Type-2. *International Journal of Computer Applications and Challenges in Networking, Intelligence and Computing Technologies*, *3*, 14–16.
39. Ephzibah, E. P. (2011). Cost effective approach on feature selection using genetic algorithms and fuzzy logic for diabetes diagnosis. *arXiv* preprint arXiv:1103.0087.
40. Levey, A. S., & Coresh, J. (2012). Chronic kidney disease. *The Lancet*, *379*(9811), 165–180.
41. Charleonnan, A., Fufaung, T., Niyomwong, T., Chokchueypattanakit, W., Suwannawach, S., & Ninchawee, N. (2016). Predictive analytics for chronic kidney disease using machine learning techniques. In *2016 Management and Innovation Technology International Conference (MITicon)* (p. MIT-80). IEEE.
42. Aljaaf, A. J., Al-Jumeily, D., Haglan, H. M., Alloghani, M., Baker, T., Hussain, A. J., & Mustafina, J. (2018). Early prediction of chronic kidney disease using machine learning supported by predictive analytics. In *2018 IEEE Congress on Evolutionary Computation (CEC)* (pp. 1–9). IEEE.
43. Ma, F., Sun, T., Liu, L., & Jing, H. (2020). Detection and diagnosis of chronic kidney disease using deep learning-based heterogeneous modified artificial neural network. *Future Generation Computer Systems*, *111*, 17–26.
44. Nithya, A., Appathurai, A., Venkatadri, N., Ramji, D. R., & Palagan, C. A. (2020). Kidney disease detection and segmentation using artificial neural network and multi-kernel k-means clustering for ultrasound images. *Measurement*, *149*, 106952.
45. Al Imran, A., Amin, M. N., & Johora, F. T. (2018). Classification of chronic kidney disease using logistic regression, feedforward neural network and wide & deep learning. In *2018 International Conference on Innovation in Engineering and Technology (ICIET)* (pp. 1–6). IEEE.
46. Navaneeth, B., & Suchetha, M. (2020). A dynamic pooling based convolutional neural network approach to detect chronic kidney disease. *Biomedical Signal Processing and Control*, *62*, 102068.
47. Brunetti, A., Cascarano, G. D., De Feudis, I., Moschetta, M., Gesualdo, L., & Bevilacqua, V. (2019). Detection and segmentation of kidneys from magnetic

resonance images in patients with autosomal dominant polycystic kidney disease. In *Intelligent Computing Theories and Application: 15th International Conference, ICIC 2019, Nanchang, China, August 3–6, 2019, Proceedings, Part II* (pp. 639–650). Springer International Publishing.

48. Vijayarani, S., & Dhayanand, S. (2015). Liver disease prediction using SVM and Naïve Bayes algorithms. *International Journal of Science, Engineering and Technology Research (IJSETR)*, *4*(4), 816–820.
49. Rajeswari, P., & Reena, G. S. (2010). Analysis of liver disorder using data mining algorithm. *Global Journal of Computer Science and Technology*, *10*(14).
50. Ansari, A. Q., & Gupta, N. K. (2011). Automated diagnosis of coronary heart disease using neuro-fuzzy integrated system. In *2011 World Congress on Information and Communication Technologies* (pp. 1379–1384). IEEE.
51. Ahsan, M. M., Mahmud, M. P., Saha, P. K., Gupta, K. D., & Siddique, Z. (2021). Effect of data scaling methods on machine learning algorithms and model performance. *Technologies*, *9*(3), 52.
52. Rubin, J., Abreu, R., Ganguli, A., Nelaturi, S., Matei, I., & Sricharan, K. (2017). Recognizing abnormal heart sounds using deep learning. *arXiv* preprint arXiv:1707.04642.
53. Miao, J. H., & Miao, K. H. (2018). Cardiotocographic diagnosis of fetal health based on multiclass morphologic pattern predictions using deep learning classification. *International Journal of Advanced Computer Science and Applications*, *9*(5).
54. Bemando, C., Miranda, E., & Aryuni, M. (2021). Machine-learning-based prediction models of coronary heart disease using naïve bayes and random forest algorithms. In *2021 International Conference on Software Engineering & Computer Systems and 4th International Conference on Computational Science and Information Management (ICSECS-ICOCSIM)* (pp. 232–237). IEEE.
55. Ram Kumar, R. P., & Polepaka, S. (2020). Performance comparison of random forest classifier and convolution neural network in predicting heart diseases. In *Proceedings of the Third International Conference on Computational Intelligence and Informatics: ICCII 2018* (pp. 683–691). Springer Singapore.
56. Singh, H., Navaneeth, N. V., & Pillai, G. N. (2019). Multisurface proximal SVM based decision trees for heart disease classification. In *TENCON 2019–2019 IEEE Region 10 Conference (TENCON)* (pp. 13–18). IEEE.
57. Desai, S. D., Giraddi, S., Narayankar, P., Pudakalakatti, N. R., & Sulegaon, S. (2019). Back-propagation neural network versus logistic regression in heart disease classification. In *Advanced Computing and Communication Technologies: Proceedings of the 11th ICACCT 2018* (pp. 133–144). Springer Singapore.
58. Patil, D. D., Singh, R. P., Thakare, V. M., & Gulve, A. K. (2018). Analysis of ECG arrhythmia for heart disease detection using SVM and cuckoo search optimized neural network. *International Journal of Engineering & Technology*, *7*(217), 27–33.
59. Liu, N., Lin, Z., Cao, J., Koh, Z., Zhang, T., Huang, G. B., Ser, W., & Ong, M. E. (2012). An intelligent scoring system and its application to cardiac arrest prediction. *IEEE Transactions on Information Technology in Biomedicine*, *16*(6), 1324–1331.
60. Yang, W., Si, Y., Wang, D., & Guo, B. (2018). Automatic recognition of arrhythmia based on principal component analysis network and linear support vector machine. *Computers in Biology and Medicine*, *101*, 22–32.

61. Borley, G., Sixsmith, J., & Church, S. (2016). How does a woman with Alzheimer's disease make sense of becoming cared for? *Dementia, 15*(6), 1405–1421.
62. Neelaveni, J., & Devasana, M. G. (2020). Alzheimer disease prediction using machine learning algorithms. In *2020 6th International Conference on Advanced Computing and Communication Systems (ICACCS)* (pp. 101–104). IEEE.
63. Collij, L. E., Heeman, F., Kuijer, J. P., Ossenkoppele, R., Benedictus, M. R., Möller, C., Verfaillie, S. C., Sanz-Arigita, E. J., van Berckel, B. N., van der Flier, W. M., & Scheltens, P. (2016). Application of machine learning to arterial spin labeling in mild cognitive impairment and Alzheimer disease. *Radiology, 281*(3), 865–875.
64. Vidushi, A. R., & Shrivastava, A. K. (2019). Diagnosis of Alzheimer disease using machine learning approaches. *International Journal of Advanced Science and Technology, 29*, 7062–7073.
65. Ahmed, S., Kim, B. C., Lee, K. H., Jung, H. Y., & Alzheimer's Disease Neuroimaging Initiative. (2020). Ensemble of ROI-based convolutional neural network classifiers for staging the Alzheimer disease spectrum from magnetic resonance imaging. *PLOS One, 15*(12), e0242712.
66. Nawaz, H., Maqsood, M., Afzal, S., Aadil, F., Mehmood, I., & Rho, S. (2021). A deep feature-based real-time system for Alzheimer disease stage detection. *Multimedia Tools and Applications, 80*, 35789–35807.

CHAPTER 2

Machine Learning-Based Diagnosis and Treatment of Cancer

ABSTRACT

Machine learning employs intelligent, rational, and bioinformatics methodologies to stimulate "discovery" from situations in which computers are utilized and to identify hidden patterns in unstructured or massive datasets. This capability is beneficial for proteomic and genomic applications that necessitate significant data analysis. Consequently, machine learning is commonly employed in the field of cancer diagnosis. The application of machine learning techniques in the field of cancer prediction is experiencing a surge in popularity. This chapter provides a discussion of the role of machine learning in enhancing our comprehension of the mechanisms behind cancer development and metastasis.

2.1 INTRODUCTION

Machine learning has the potential to significantly transform cancer research and therapy. The use of machine learning in the healthcare sector has been made possible by the integration of electronic medical records and other ways of digitizing patient data. By implementing this modification, hitherto unattainable knowledge regarding patient care can be extracted from vast datasets at an unprecedented rate. When confronted with a novel clinical setting, healthcare professionals commonly consult established practices and recommendations within their respective specialties. By employing machine learning techniques, this procedure becomes

Machine Learning in Healthcare: Advances and Future Prospects. Rishabha Malviya, Niranjan Kaushik, Tamanna Rai, M. P. Saraswathy, and Rajendra Awasthi (Authors)

more rigorous, enabling computers to generate personalized predictions by analyzing a diverse set of patient data [1]. The use of this data for the purpose of policymaking can facilitate the establishment of standards and the identification of vulnerabilities based on data-driven approaches. The enhanced level of clarity facilitates the provision of more precise medical treatment tailored to the individual circumstances of each patient. As a result, machine learning approaches have rapidly disseminated across the medical field. This algorithm efficiently identifies patterns and correlations within intricate datasets, enabling the prediction of cancer outcomes [2]. This exhibits a common requirement for the integration of data originating from diverse sources, including medical records and DNA sequences. Nevertheless, it was observed that a substantial number of the studies examined in our analysis neglected to assess the validity of their models through a rigorous comparison with empirical data from the real world. This implies that the use of machine learning techniques could potentially enhance our ability to predict the likelihood of a patient developing, encountering, or surviving cancer. According to the findings described by Aha [3], the implementation of machine learning techniques has resulted in a notable enhancement of approximately 15–20% in the precision of cancer prediction outcomes. The evaluation focused exclusively on the assessment of cancer diagnosis and diagnostic studies conducted using machine learning simulations.

2.2 MACHINE LEARNING TECHNIQUES

The field of machine learning establishes a connection between the process of data sampling and the process of making inferences in the context of AI [4]. The initial stage of any learning procedure involves employing an available dataset to generate approximations regarding the hidden interconnections within the system. Subsequently, these approximations are used to forecast the forthcoming outcomes of the system [5, 6]. The application of machine learning has demonstrated potential in the field of biological diagnostics, namely in the identification of suitable generalizations through the exploration of an n-dimensional space for a specific set of biological fluids [7]. Supervised learning, which is one of the predominant machine learning methodologies, aims to predict an established outcome, such as the detection of cancer, the lifespan of a patient, or the efficacy of therapeutic interventions. Unsupervised learning is a valuable approach for discerning patterns and subgroups within data in situations where a definitive conclusion cannot be readily drawn. Exploratory studies are commonly conducted. Reinforcement learning (RL), as a form of machine learning, is particularly suited for the sequential decision-making process where a strategy needs to be acquired through data analysis. This approach has been found to be effective in determining the optimal cancer treatments with the highest likelihood of success [8].

By utilizing the testing set to determine the model's predicted accuracy, one can gain insights into the extent of generalization errors. To obtain precise and reliable insights into the predictive capabilities of the 209 model, it is imperative to employ large and independent training and testing datasets, including appropriately labeled testing data. There are several methodologies for evaluating the effectiveness of a classifier, with four prominent approaches being bootstrapping, the holdout technique, cross-validation, and random sampling. During the holdout procedure, the data samples are partitioned into two distinct subsets: the training set and the evaluation set [9].

2.3 MACHINE LEARNING APPROACHES

Various machine learning algorithms, including artificial neural networks (ANNs), support vector machines (SVMs), linear models, and decision trees (DTs), can be utilized once the data has been appropriately prepared and the specific learning objective has been defined. This section centers on machine learning methodologies that have been frequently used across the scientific community with the aim of generating predictions and prognostications for cancer. In order to anticipate the development of cancer and evaluate the consequences of the disease, a thorough compilation is shown below, encompassing the frequently utilized machine learning approaches, the integrated data types, and the assessment procedures used to gauge the overall effectiveness of these systems.

2.3.1 ARTIFICIAL NEURAL NETWORK (ANN)

The utilization of ANNs enables the effective resolution of a diverse array of classification and pattern recognition problems. Due to their rigorous training, ANNs possess the ability to process and integrate various inputs, ultimately generating a singular output. The utilization of several hidden layers is employed in order to mathematically represent the connections within the brain. Although ANNs are widely recognized as the prevailing method for numerous classification problems [10], they do possess certain limitations. The typical layered structure employed in this context is both time-consuming and potentially characterized by inefficiency that may lead to risky outcomes. Furthermore, the phrase "black box" is frequently employed to characterize this kind of functioning. In certain instances, such as when attempting to ascertain the reasons behind the malfunctioning of an ANN, the mechanism by which it achieves categorization can be challenging to comprehend. An ANN, as depicted in Figure 2.1, is comprised of a set of interconnected nodes.

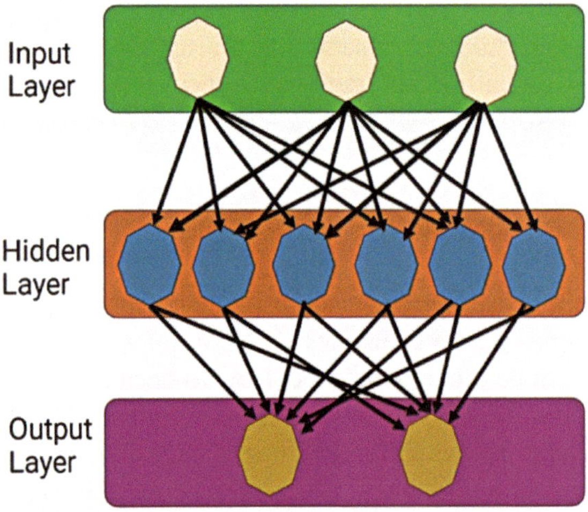

FIGURE 2.1 Schematic presentation of an artificial neural network. The arrow serves to establish a connection between the output of one node and the input of another.

2.3.2 *SUPPORT VECTOR MACHINES (SVMS)*

Support vector machines (SVMs) present an innovative strategy for utilizing machine learning techniques in the context of predicting the probability of cancer incidence. SVM use a hyperplane to split the input vector into two distinct classes by projecting it into a feature space of higher dimensionality. This objective is achieved by optimizing the marginal distance between the selected hyperplane and the instances located at the boundary. Once a high level of generalizability has been attained, the resulting classifier can be efficiently utilized for the categorization of novel samples. It is important to note that SVMs can provide probabilistic outputs [11]. Figure 2.2 depicts the use of a SVM in differentiating between benign and malignant malignancies by incorporating tumor size and patient age as distinguishing factors. The hyperplane serves as a discriminant boundary for distinguishing between the two groups. The identification of any incorrect classifications produced by the technique is possible due to the existence of a decision boundary.

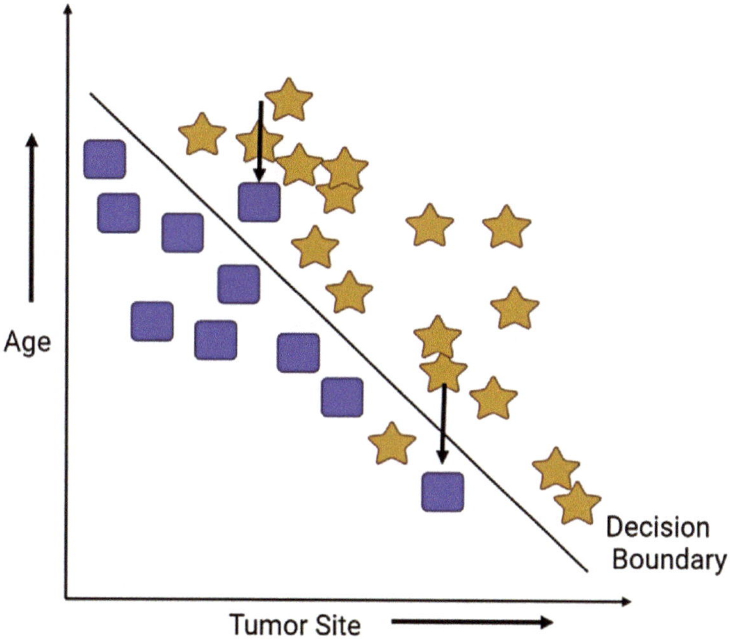

FIGURE 2.2 A simplified representation of the input data being classified by a linear support vector machine model. Tumors are classified based on both the tumor size and patient age. The arrow indicates the presence of malignant tumors that have been erroneously labeled.

2.3.3 LINEAR MODELS

Linear models establish a relationship between an independent variable and a dependent variable. Linear regression is a statistical method used to estimate the coefficients of features. Subsequently, a set of weights ($\beta_0 + \beta_1 x_1 + \ldots + \beta pxp$) is applied, where x_1 to xp represent independent factors that include the patient's features, to forecast an observation. The underlying assumption of linear regression is that the features exhibit additivity, meaning that the effect of each feature on the outcome is independent of the other characteristics. Additionally, linear regression requires that the relationship between the feature values and the outcome follows a linear pattern.

Logistic regression (LR) and Cox regression, along with other statistical models, also assume an additive relationship between the features. However, these models adjust the linear function based on the specific prediction task at hand. Linear techniques are commonly used by modelers because of their simplicity and interpretability. These models serve as the foundation for healthcare risk evaluations and predictive models.

However, numerous findings exhibit nonlinearity. For example, the impact of tumor size on cancer recurrence demonstrates varying patterns across different age groups. The complex relationships among several factors often elude comprehension with a linear approach. Nonlinearities can be effectively addressed through the use of interaction factors, such as a derived feature that incorporates both age and tumor size, thereby explaining their combined impact. Due to the time-consuming nature of exploring every conceivable modification of pairs or larger groupings of variables, this process is often conducted in an ad hoc manner [13].

2.3.4 DECISION TREES (DTS)

Decision trees (DTs) employ a classification tree structure in which the leaves correspond to judgments and the nodes correspond to input elements. DTs are a widely used and well-established machine learning categorization technique. DTs are simple and quick to learn. The estimation of a sample's category can be achieved by utilizing the branches of a classification tree. The unique architectural features of the subject under consideration provide enough justification to render their judgments highly attractive. Figure 2.3 illustrates the components and restrictions of the DT system.

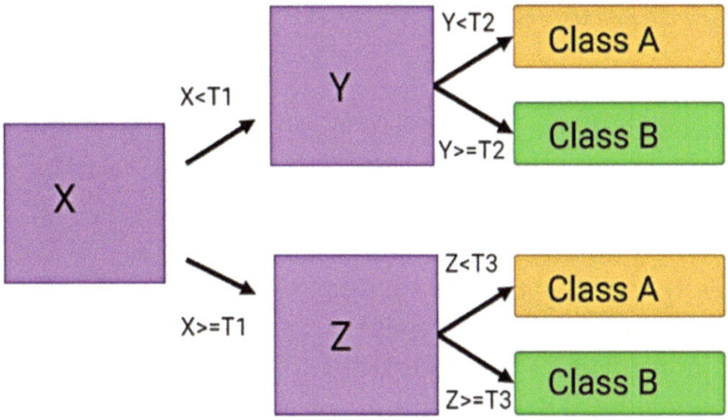

FIGURE 2.3 Diagram illustrating a decision tree in a tree-like format. The classification rule consists of independent variables *X*, *Y*, and *Z*, with possible outcomes being class A and class B. The rule also includes thresholds T(1–3) for accurately assigning each variable to a class label.

2.3.5 ENSEMBLE MODELS

Integration of gradient-boosted machines and random forests (RFs) is one potential approach to optimizing the decision-making tree structure [14]. These methodologies include the generation of a significant number of DTs, which are subsequently utilized for prediction purposes. RFs generate diverse models by training each tree with a random subset of features and data. The final forecast combines the prognostications of the trees. Gradient-boosted machines employ an iterative training process wherein individual trees are trained by assigning weights to data points based on the inaccuracies of earlier trees. Error-correcting approaches have been found to outperform RFs. Ensemble techniques, by aggregating several trees, do not establish a clear linkage between input features and the final prediction. Consequently, these models provide greater challenges in terms of interpretability compared to linear models with coefficients or DTs with feature partitions. The absence of clear understanding poses a significant challenge in situations where users depend on straightforward comprehension to adopt new software [12, 13]. In such cases, Shapley frameworks offer a more comprehensive understanding through their ability to generate broader insights from measures of model feature relevance. Additive explanations are commonly employed to acquire valuable insights [15, 16].

2.4 MACHINE LEARNING APPLICATIONS IN CANCER

The diagnosis of cancer necessitates the utilization of early identification techniques such as gene expression analysis, radiography, histology, or a combination thereof. Since the early 2000s, machine learning techniques have been utilized to identify cancer biomarkers by analyzing gene expression profiles [17–20]. The field of computer vision has advanced to a stage where it is now feasible to analyze and interpret raw images to make a diagnosis. Mammograms have consistently emerged as a very efficacious method for cancer detection. The detection of breast cancer is the prominent focus of study within this domain. There have been studies in this area since 1995, and recent improvements in mammography-based identification are the most important ones [21–23]. Various methodologies utilizing computed tomography (CT) scans have been devised in similar ways to facilitate the diagnosis of lung cancer [24]. Hu et al. conducted a comprehensive analysis of the diagnostic imaging software currently available [25]. The potential applications of image-based diagnosis in the field of histology have also been reported [26]. Convolutional neural networks (CNNs) have been employed in several diagnostic tasks, including pathological outcomes. Notable instances include the identification of prostate cancer [27], bladder cancer [28], and lymph node dissection for the diagnosis of breast cancer [27, 29].

The ability of machine learning to identify patient abnormalities over an extended period implies that machine learning may have applications in the early detection of cancer. Even though early cancer identification is vital, it is challenging to accomplish due to the complexity and individuality of the signs that point to the beginning of cancer [30]. AI technologies have been utilized to make predictions for future diagnoses of breast cancer in tissue growth and repair through mammography [31], as well as lung cancer through CT scans [32]. Gene expression data has been utilized in a number of studies to determine cancer susceptibility, and data from electronic medical records has been used to predict the incidence of pancreatic cancer in high-risk people [33]. These early detection systems offer an early indication of cancer existence, which may help guide policy and practices related to cancer screening. Most importantly, they enable the potential for earlier intervention, which has the potential to improve patient outcomes [34].

2.5 CONCLUSION

The field of oncology has great expectations for the potential of machine learning. The utilization of this tool can contribute to the facilitation of diagnosis and early intervention, as well as aid in the identification of high-risk populations and the prediction of prognosis. The utilization of data-driven methodologies has the potential to enhance our understanding of cancer and its impact on individuals, particularly given the substantial volume of patient data currently available. The field of cancer treatment stands to undergo a considerable transformation through the integration of machine learning techniques. However, the realization of this potential advancement is dependent upon the successful navigation of substantial technological and organizational challenges.

KEYWORDS

- **artificial neural networks**
- **cancer**
- **decision trees**
- **linear models**
- **machine learning**
- **support vector machines**

REFERENCES

1. Suhel, S. F., Shukla, V. K., Vyas, S., & Mishra, V. P. (2020). Conversation to automation in banking through chatbot using artificial machine intelligence language. In *2020 8th International Conference on Reliability, Infocom Technologies and Optimization (Trends and Future Directions) (ICRITO)* (pp. 611–618). IEEE.
2. Polley, M. Y., Freidlin, B., Korn, E. L., Conley, B. A., Abrams, J. S., & McShane, L. M. (2013). Statistical and practical considerations for clinical evaluation of predictive biomarkers. *Journal of the National Cancer Institute*, *105*(22), 1677–1683.
3. Aha, D. W. (1992). Tolerating noisy, irrelevant, and novel attributes in instance-based learning algorithms. *International Journal of Man-Machine Studies*, *36*(2), 267–287.

4. Ji, S., Pan, S., Cambria, E., Marttinen, P., & Philip, S. Y. (2021). A survey on knowledge graphs: Representation, acquisition, and applications. *IEEE Transactions on Neural Networks and Learning Systems*, *33*(2), 494–514.
5. Mitchell, T. M. (2006). *The Discipline of Machine Learning*. Pittsburgh: Carnegie Mellon University, School of Computer Science, Machine Learning Department.
6. Bandara, K., Bergmeir, C., & Smyl, S. (2020). Forecasting across time series databases using recurrent neural networks on groups of similar series: A clustering approach. *Expert Systems with Applications*, *140*, 112896.
7. Niknejad, A., & Petrovic, D. (2013). Introduction to computational intelligence techniques and areas of their applications in medicine. *Med Appl Artif Intell*, *51*, 201.
8. Landers, M., & Doryab, A. (2023). Deep reinforcement learning verification: A survey. *ACM Computing Surveys*.
9. Padmanabhan, R., Meskin, N., & Haddad, W. M. (2017). Reinforcement learning-based control of drug dosing for cancer chemotherapy treatment. *Mathematical Biosciences*, *293*, 11–20.
10. Picano, E. (2004). Informed consent and communication of risk from radiological and nuclear medicine examinations: How to escape from a communication inferno. *BMJ*, *329*(7470), 849–851.
11. Alexander, D. D., Mink, P. J., Cushing, C. A., & Sceurman, B. (2010). A review and meta-analysis of prospective studies of red and processed meat intake and prostate cancer. *Nutrition Journal*, *9*, 1–7.
12. Breiman, L. (2001). Random forests. *Machine Learning*, *45*, 5–32.
13. Schneider, A., Hommel, G., & Blettner, M. (2010). Linear regression analysis: Part 14 of a series on evaluation of scientific publications. *Deutsches Ärzteblatt International*, *107*(44), 776.
14. Chandra, B., & Varghese, P. P. (2009). Fuzzifying Gini index-based decision trees. *Expert Systems with Applications*, *36*(4), 8549–8559.
15. Lundberg, S. M., & Lee, S. I. (2017). A unified approach to interpreting model predictions. *Advances in Neural Information Processing Systems*, *30*.
16. Lundberg, S. M., Erion, G., Chen, H., DeGrave, A., Prutkin, J. M., Nair, B., Katz, R., Himmelfarb, J., Bansal, N., & Lee, S. I. (2020). From local explanations to global understanding with explainable AI for trees. *Nature Machine Intelligence*, *2*(1), 56–67.
17. Jones, W., Alasoo, K., Fishman, D., & Parts, L. (2017). Computational biology: Deep learning. *Emerging Topics in Life Sciences*, *1*(3), 257–274.
18. Hwang, K. B., Cho, D. Y., Park, S. W., Kim, S. D., & Zhang, B. T. (2002). Applying machine learning techniques to analysis of gene expression data: Cancer diagnosis. *Methods of Microarray Data Analysis: Papers from CAMDA'00*, 167–182.
19. Danaee, P., Ghaeini, R., & Hendrix, D. A. (2017). A deep learning approach for cancer detection and relevant gene identification. In *Pacific Symposium on Biocomputing 2017* (pp. 219–229).
20. Ye, Q. H., Qin, L. X., Forgues, M., He, P., Kim, J. W., Peng, A. C., Simon, R., Li, Y., Robles, A. I., Chen, Y., & Ma, Z. C. (2003). Predicting hepatitis B virus–positive metastatic hepatocellular carcinomas using gene expression profiling and supervised machine learning. *Nature Medicine*, *9*(4), 416–423.

21. Wolberg, W. H., Street, W. N., & Mangasarian, O. L. (1995). Image analysis and machine learning applied to breast cancer diagnosis and prognosis. *Analytical and Quantitative Cytology and Histology*, *17*(2), 77–87.
22. Shen, L., Margolies, L. R., Rothstein, J. H., Fluder, E., McBride, R., & Sieh, W. (2019). Deep learning to improve breast cancer detection on screening mammography. *Scientific Reports*, *9*(1), 12495.
23. Karthik, R., Menaka, R., Kathiresan, G. S., Anirudh, M., & Nagharjun, M. (2022). Gaussian dropout based stacked ensemble CNN for classification of breast tumor in ultrasound images. *IRBM*, *43*(6), 715–733.
24. Sun, W., Zheng, B., & Qian, W. (2017). Automatic feature learning using multichannel ROI based on deep structured algorithms for computerized lung cancer diagnosis. *Computers in Biology and Medicine*, *89*, 530–539.
25. Hu, Z., Tang, J., Wang, Z., Zhang, K., Zhang, L., & Sun, Q. (2018). Deep learning for image-based cancer detection and diagnosis—A survey. *Pattern Recognition*, *83*, 134–149.
26. Madabhushi, A. (2009). Digital pathology image analysis: Opportunities and challenges. *Imaging in Medicine*, *1*(1), 7.
27. Song, Y., Zhang, Y. D., Yan, X., Liu, H., Zhou, M., Hu, B., & Yang, G. (2018). Computer-aided diagnosis of prostate cancer using a deep convolutional neural network from multiparametric MRI. *Journal of Magnetic Resonance Imaging*, *48*(6), 1570–1577.
28. Zhang, Z., Chen, P., McGough, M., Xing, F., Wang, C., Bui, M., Xie, Y., Sapkota, M., Cui, L., Dhillon, J., & Ahmad, N. (2019). Pathologist-level interpretable whole-slide cancer diagnosis with deep learning. *Nature Machine Intelligence*, *1*(5), 236–245.
29. Liu, Y., Gadepalli, K., Norouzi, M., Dahl, G. E., Kohlberger, T., Boyko, A., Venugopalan, S., Timofeev, A., Nelson, P. Q., Corrado, G. S., & Hipp, J. D. (2017). Detecting cancer metastases on gigapixel pathology images. *arXiv Preprint arXiv:1703.02442*.
30. Burke, H. B., Goodman, P. H., Rosen, D. B., Henson, D. E., Weinstein, J. N., Harrell Jr, F. E., Marks, J. R., Winchester, D. P., & Bostwick, D. G. (1997). Artificial neural networks improve the accuracy of cancer survival prediction. *Cancer*, *79*(4), 857–862.
31. Yala, A., Lehman, C., Schuster, T., Portnoi, T., & Barzilay, R. (2019). A deep learning mammography-based model for improved breast cancer risk prediction. *Radiology*, *292*(1), 60–66.
32. Ciompi, F., Chung, K., Van Riel, S. J., Setio, A. A., Gerke, P. K., Jacobs, C., Scholten, E. T., Schaefer-Prokop, C., Wille, M. M., Marchiano, A., & Pastorino, U. (2017). Towards automatic pulmonary nodule management in lung cancer screening with deep learning. *Scientific Reports*, *7*(1), 46479.
33. Kim, B. J., & Kim, S. H. (2018). Prediction of inherited genomic susceptibility to 20 common cancer types by a supervised machine-learning method. *Proceedings of the National Academy of Sciences*, *115*(6), 1322–1327.
34. Chen, X., Gole, J., Gore, A., He, Q., Lu, M., Min, J., Yuan, Z., Yang, X., Jiang, Y., Zhang, T., & Suo, C. (2020). Non-invasive early detection of cancer four years before conventional diagnosis using a blood test. *Nature Communications*, *11*(1), 3475.

CHAPTER 3

Machine Learning-Based Detection and Management of Cardiovascular Diseases

ABSTRACT

The discipline of artificial intelligence (AI) is experiencing rapid expansion and is gaining significant attention across various domains, including healthcare. Consequently, the integration of AI in the healthcare sector has facilitated advancements in early detection, prognostication, and mitigation of many ailments, encompassing cardiovascular disorders. Heart disease is considered one of the most fatal diseases. The prediction of cardiovascular disease using clinical data is difficult. This chapter explores the use of artificial neural networks (ANNs), support vector machines (SVMs), decision trees (DTs), random forests (RFs), and K-nearest neighbors (KNNs) in the diagnosis of cardiac diseases. The use of machine learning is expected to play a pivotal role in facilitating the monitoring, diagnosis, and prediction of cardiovascular disease and several other health issues. This advancement would enable healthcare professionals to make more informed decisions regarding treatment options.

3.1 INTRODUCTION

The heart plays a crucial role as an essential organ in the human body. The circulatory system facilitates the transportation of blood throughout the body via a network of arteries. The circulatory system serves a crucial role in the delivery of nutrients to the various cells and tissues throughout the body [1]. Malfunctions in cardiac function can have profound implications

Machine Learning in Healthcare: Advances and Future Prospects. Rishabha Malviya, Niranjan Kaushik, Tamanna Rai, M. P. Saraswathy, and Rajendra Awasthi (Authors)

for overall health, potentially resulting in death. During the latter part of the 19th century, significant advancements were made in the field of cardiovascular care with the development of the electrocardiogram (ECG). Since its inception, technology has played a crucial role in facilitating the rapid evolution of this discipline [2]. In anticipation of the integration of artificial intelligence (AI) and machine learning, the aforementioned sector has extensively employed technology within clinical settings and promptly embraced guideline-directed medical practice as a means to enhance patient outcomes. The diagnosis and management of cardiovascular disease in clinical practice commonly involve the utilization of many data sets [3].

Clinicians have various methods for communicating clinical data, such as a patient's medical record, laboratory findings, imaging modalities, physical examination, or angiography [4]. Cardiologists are currently facing the challenge of performing increasingly complex assessments due to an increasing number of data-driven technologies, including mobile telemetry devices, wearable, and implanted recording devices, statistics derived from electronic health records (EHRs), and patient-generated health data [5]. The process of clinical decision-making is subject to various factors beyond the mere consideration of factual information and personal experience [6]. In the field of cardiovascular medicine, a comparable imperative exists, as in other fields, to maximize patient care by concurrently reducing costs and enhancing productivity. To administer such personalized treatment, a substantial amount of data is necessary. However, the dynamic nature of this data presents challenges in its utilization without a comprehensive clinical decision support tool. The absence of cognitive computing would result in the continued presence of issues like overutilization and insufficient patient care, which, in turn, would impact readmission and death rates within our region [7].

3.2 CARDIOVASCULAR DISEASES

Cardiovascular disease comprises a wide range of diseases affecting the heart and blood vessels. Both angina and heart attacks are classified as cardiovascular disorders, specifically resulting from coronary artery disease. Coronary heart disease is attributed to the presence of plaque within the coronary arteries. Atherosclerosis is a pathological condition characterized by the accumulation of plaque within the arterial walls. The accumulation of plaque occurs gradually over many years. Over time, this plaque can undergo either hardening or crumbling, resulting in its potential breakage. The coronary arteries undergo a process of arterial stiffening and constriction due to the accumulation of plaque, thereby impeding the delivery of oxygenated blood to the cardiac muscle. If the plaque becomes disrupted, a thrombus can develop on its surface. The occurrence of a total obstruction of a coronary artery by a substantial blood clot is relatively uncommon. The accumulation of plaque leads to the progressive stiffening and narrowing of arterial walls. Rapid restoration of blood flow is necessary to prevent myocardial necrosis. If left untreated, a heart attack has the potential to result in death. Myocardial infarctions constitute a prominent contributor to global mortality rates. The subsequent manifestations are indicative of a myocardial infarction.

Chest pain is the most common symptom of a heart attack. Clogged arteries can cause chest pain, tightness, or pressure. Heart attacks and clogged arteries can cause tightness, chest pain, and pressure. In addition to stroke and heart attack, there are several other prevalent cardiovascular problems, namely heart failure, hypertension, rheumatic heart disease, cardiomyopathy, cardiac arrhythmia, congenital heart disease, valvular heart disease, aortic aneurysms, peripheral artery disease, and venous thrombosis. Smoking, inadequate dietary habits, and a lack of physical activity are recognized as risk factors associated with the development of cardiovascular disease. Early detection of cardiac disease allows for the possibility of effective treatment. The timely identification of a phenomenon is of utmost importance. Gaining knowledge about the underlying factors contributing to the development of heart disease is important for implementing effective preventive measures. In cardiac patients, the chest pain often spreads to the arms, especially the left side.

3.3 CARDIOVASCULAR EPIDEMIOLOGY

Each year, 17.5 million people lose their lives due to cardiovascular disease. Middle- and low-income countries account for 75% of cardiovascular disease mortality. Heart disease or stroke kills about 80% of people with cardiovascular disease [8]. Each year, an increasing number of people in India are diagnosed with cardiovascular disease. The number of individuals in India who have received a diagnosis of heart disease has exceeded 30 million. Over 2 million open-heart surgeries are carried out in India annually. The number of patients requiring coronary surgery has been increasing by 20–30% annually [9].

3.4 ALGORITHMS FOR MACHINE LEARNING

Numerous data mining algorithms have been developed because of extensive research in the field of data mining. These strategies can be directly applied to a dataset to construct models or extract valuable insights. In the area of data mining, commonly employed techniques include decision trees (DTs), naive bayes (NB), k-means, artificial neural networks (ANNs), and similar methodologies. The subsections will provide additional analysis on this subject.

3.4.1 RANDOM FOREST (RF) CLASSIFICATION AND REGRESSION

The random forest (RF) algorithm is a popular technique in machine learning that utilizes ensemble learning to perform both classification and regression tasks. In the training phase, several separate DTs are formed, with each tree generating a singular prediction regarding the target class. The final output is the most frequently taken class in the field of forecasting. The objective is to find a point of agreement by employing the method of averaging to mitigate the effects of both significant disparities and pronounced bias. Both R and Python offer libraries that provide robust support for this methodology.

3.4.2 NAIVE BAYES (NB) CLASSIFIER (SUPERVISED ALGORITHM)

The Bayes' theorem-based categorization is simple. The assumption exhibits a simplistic notion of independence. Bayes' theorem is an algorithm used to calculate the likelihood of an event occurring, given prior knowledge or information. There is no discernible relationship between the predictors. The overall success of the entity can be attributed to its various attributes. As it does not utilize Bayesian methods, it aligns with the NB model in terms of consistency. NB classifiers are commonly employed in diverse practical contexts that encompass both sophisticated and pragmatic applications [10].

$$P\left(\frac{X}{Y}\right)=\frac{P\left(\frac{Y}{X}\right)\times P(X)}{P(Y)}$$

Within this particular framework, the notation *P(X/Y)* represents the posterior probability; *P(X)* denotes the class prior probability; *P(Y)* signifies the predictor prior probability; and *P(Y/X)* corresponds to the likelihood, or probability, of the predictor. The NB algorithm is a classification technique that is known for its simplicity, ease of implementation, and computational efficiency. It is particularly effective in handling non-linear and complex datasets. However, its reliance on assumptions and class-conditional independence results in a decrease in accuracy. The NB model demonstrated an improved accuracy of 84.1584% when a feature selection technique known as SVMRFE was employed to choose the top 10 predictors [11]. The study conducted on the Cleveland dataset utilized all 13 parameters and determined that an accuracy rate of 83.49% was deemed satisfactory [12].

3.4.3 K-NEAREST NEIGHBOR (KNN)

The K-nearest neighbors (KNNs) algorithm is commonly employed as a method for supervised classification. This technique utilizes a nearest-neighbor classification approach for categorizing items into separate groups. The strategy employed in this case is known as instance-based learning (IBL). The Euclidean distance metric is used to determine the degree of separation between the two attributes [3]. The method uses designated points to indicate a different point. The KNNs algorithm can address missing data by employing a clustering approach to identify and utilize comparable data. Once missing values are filled, prediction methods can be applied to the data set. The integration of multiple algorithms enhances the precision of the results. The KNNs technique does not necessitate the establishment of models or the application of assumptions. This algorithm is utilized in regression, classification, and search applications. It is important to note that the presence of noisy and irrelevant information can significantly affect the performance of the KNNs algorithm. In their study, Pouriyeh et al. reported an accuracy rate of 83.16% when using a K-value of 9 [12].

3.4.4 ARTIFICIAL NEURAL NETWORK (ANN)

An artificial neural network (ANN) is a type of computing model that closely resembles the organization and behavior of natural neural networks. A neural network can learn and adapt based on the input and output it receives at each level of the network, and the data passing through the network can modify its structure. Nonlinear statistical data modeling is an area of data science that makes use of ANNs. The objective of this particular subfield is to either formulate or ascertain complex relationships between input variables and output variables. ANNs consist of layers that are connected. ANNs are being utilized to progress the current state of the art in data processing. These networks have a simple mathematical architecture.

3.4.5 DECISION TREE (DT)

Decision trees (DTs) are a type of machine learning algorithm that can be used to categorize data, both numerical and categorical. DTs are a type of data structure that establishes hierarchical relationships. DTs are frequently employed in medical datasets due to their simplicity in both design and interpretation. Tree-shaped graphs simplify the process of data processing. A DT analysis is conducted at three distinct nodes. The root node serves as the central point around which all other nodes spin. At a node within a data structure, referred to as an "internal node," many properties are managed. The test findings are represented by the leaf nodes. The algorithm in question integrates data into cohesive clusters to facilitate comprehension and analysis. The calculation of attribute entropy is initially performed to classify the data that possess the most accurate predictors.

$$\text{Entropy}(S) = \sum_{i=1}^{c} -\text{Pi} \log_2 \text{Pi},$$

$$\text{Gain}(S, A) = \text{Entropy}(S) - \sum_{v \in \text{Value}(A)} \frac{|Sv|}{|S|} \text{Entropy}(Sv)$$

The results provided in this study exhibit improved readability and comprehensibility [13]. Due to the utilization of a tree-based graphical representation to analyze the dataset, this approach demonstrates superior precision compared to its counterparts. However, it is important to note that judgements typically focus on a single property, potentially leading to an over classification of the data. The DT model developed by Chauhan et al. demonstrates an accuracy rate of 71.43%. However, the achieved accuracy is notably lower at 42.895% [14].

3.4.6 FUZZY LOGIC

In this form of multi-valued logic, the variables can assume truth values that span the range of positive or negative real numbers inside the interval of 0 to 1. Fuzzy logic has the potential to offer valuable contributions across various disciplines, ranging from control theory to the field of AI. The phenomenon of partial truth, wherein the degree of truthfulness can vary from completely true to completely false, is frequently examined through the use of fuzzy logic. The advancement of neuro-fuzzy systems can be facilitated using diverse hybrid methodologies within the area of soft computing, wherein one such strategy involves the combination of fuzzy logic with neuro-computing.

3.4.7 ASSOCIATION RULES

Association rules enable data warehouses to discover associations between apparently unrelated data by utilizing if/then statements. The statement contains an if-then clause, namely the "then" part. The identification of recurring if/then patterns within a dataset leads to the establishment of association rules. The determination of the most significant connections is dependent on support and confidence measures. The confidence metric quantifies the dependability of the if/then expressions, whereas the support metric quantifies their occurrence frequency inside the database. The utilization of association rules in data mining has the potential to forecast customer behavior. Programmers employ association rules to develop machine learning systems [15].

3.4.8 CLASSIFICATION AND REGRESSION TREES

Classification and regression trees are a type of DT algorithm used for classifying categorical target variables. Regression trees are a type of predictive model that is used to estimate and forecast continuous target variables. The technique known as classification and regression trees is comprised of a sequence of questions that ascertain the subsequent questions, if applicable. These questions constitute a hierarchical structure resembling a tree, wherein the terminal nodes signify the lack of any further inquiry.

3.5 DATA MINING TOOLS

The mining methods can be readily implemented using data mining tools. To facilitate accessibility for researchers, most of such software applications are made available as open source without any associated costs. Their design makes them easy to use. Various platforms and software tools, such as Waikato Environment for Knowledge Analysis (WEKA), RapidMiner, Tanagra, MATLAB, and so on, are extensively employed within the domain of data mining.

3.5.1 WAIKATO ENVIRONMENT FOR KNOWLEDGE ACQUISITION

The Waikato University-developed computer program aims to extract useful information from unstructured data. The WEKA software package possesses the capability to undertake many tasks, including pre-processing, classification, clustering, regression analysis, visualization, and feature selection. This program uses computer software to acquire knowledge in the field of machine learning and identify common patterns and trends. The WEKA software, which was first implemented in the C programming language, has undergone a rewriting process in Java to ensure compatibility with a wide range of computer systems. The efficient graphical user interface facilitates rapid setup and utilization [16].

3.5.2 RAPIDMINER

RapidMiner, formerly referred to as YALE, is a software tool that facilitates various data mining and machine learning tasks, including ETL (extract, transform, load), data preparation and visualization, modeling, assessment, and deployment. The RapidMiner software utilizes the Java programming language. Text mining, media processing, feature engineering, data stream mining, and others are viable.

3.5.3 C PROGRAMMING LANGUAGE

The programming language C was developed by Dennis M. Ritchie during the early 1970s at Bell Labs for the Unix operating system. The initial objective of system software was to fulfill its designated functions. The programming language C is well-suited for the development of firmware and portable applications.

3.5.4 JAVA PROGRAMMING LANGUAGE

The software was initially developed by Sun Microsystems, but it is presently in the ownership of Oracle. It is extensively used to create and distribute web content. Java is an object-oriented programming language that shares many features with C^{++} but is simpler and less prone to programming errors. Java is an excellent programming language for use on the internet. Java applets can be retrieved from the web server and subsequently executed within a web browser that is compatible with Java.

3.5.6 APACHE MAHOUT

The development of open-source implementations of scalable machine learning algorithms is the major objective of the Apache Mahout project, which is being supervised by the Apache Software Foundation. These techniques can be applied to classification, clustering, and collaborative filtering. Apache Hadoop, an open-source platform based on Java, enables distributed computing for the purpose of processing and storing large volumes of data. The Apache Software Foundation provides support for it as an integral component of the Apache project.

3.5.7 ORANGE

This software package comprises a collection of tools designed for the purpose of visually analyzing and examining information, as well as for using machine learning algorithms and conducting data mining operations.

This is a Python library that has the capability to be used in an interactive manner.

3.5.8 MATLAB

Matrix Lab is an acronym for the phrase matrix laboratory. This software exhibits compatibility with a diverse range of numerical computing paradigms. This is a programming language belonging to the fourth generation of computer programming. MATLAB facilitates the execution of matrix operations, data, and function plotting, algorithm creation, user interface design, and inter-program communication with languages such as Java, C^{++}, C#, Fortran, and Python [17].

3.5.9 TANAGRA

Tanagra data mining projects often employ it due to its low cost and non-commercial availability. Exploratory analysis, statistical learning, and machine learning are all recommended data mining methodologies. Among the paradigms that Tanagra uses are clustering, association rules, parametric, and nonparametric statistics, factorial analysis, feature selection, and model generation.

3.6 INVESTIGATIONS BASED ON THE USE OF COMPUTERS TO FORECAST CARDIAC DISEASES

Gudadhe et al. introduced a decision-support system for cardiovascular disease classification. The system mainly utilized ANNs and support vector machine (SVM) techniques. A heart disease diagnostic decision support system was built using a three-layer multilayer perceptron (MP) neural network. This layered perceptron network was trained using the computationally efficient back-propagation approach. The results demonstrated the efficacy of a MP neural network trained by back-propagation [18].

Ordonez used association rule mining and the train-test strategy to analyze a set of data to predict heart disease. The biggest issue with association rule mining is that it produces many rules, the majority of which are irrelevant to the healthcare industry. In addition, most of the time, association rules are extracted using the entire set of data without being tested on a subset. The author has come up with a solution to this issue in the form of an algorithm that employs search constraints to reduce the size of the ruleset. The system initially analyzes the training data to identify potential association rules, which are subsequently evaluated against an independent test data set to assess their accuracy. Subsequently, Ordonez conducted a thorough and optimistic evaluation of the medical significance of the newly implemented restrictions. By implementing search limits and test set validation, the number of association rules is significantly diminished while simultaneously attaining a high level of prediction accuracy. These recommendations are an invaluable resource for healthcare professionals [19].

Bhatt et al. (n.d.) devised a model capable of accurately forecasting cardiovascular disorders to mitigate the mortality associated with such conditions using a novel approach to k-mode clustering, utilizing the Huang initialization method, which has the potential to enhance classification accuracy. The models employed include RF, DT classifier, MP, and XGBoost (XGB). The GridSearchCV function was employed to optimize the parameters of the applied model. The models were trained using an 80:20 data split (the DT model achieved an accuracy of 86.37% with cross-validation and 86.53% without cross-validation). The XGB model achieved an accuracy of 86.87% with cross-validation and 87.02% without cross-validation. The RF model achieved an accuracy of 87.05% with cross-validation and 86.92% without cross-validation. The MP model achieved an accuracy of 87.28% with cross-validation and 86.94% without cross-validation. The study concluded that the MP, when combined with cross-validation, has exhibited superior accuracy (a peak accuracy of 87.28%) compared to all other algorithms [20]. Apache Mahout, developed by the Apache Software Foundation, is freely available open-source software designed to execute machine learning algorithms in a distributed or scalable environment. The Cleveland Heart Database is an openly accessible online repository containing a collection of data encompassing 13 different variables. Three methodologies, namely a neural network, NB, and a DT, have been employed to extract the underlying patterns. The primary goal of this system is to enhance the precision of prediction models, risk assessment tools, and feature extraction mechanisms concerning a broader range of clinical hazards.

To diagnose cardiac issues, supervised machine learning classification strategies have been explored. The Tanagra tool is used to organize the data, which is then tested using 10-fold cross-validation, and the findings are compared. Tanagra is open-source software that is freely available for use in educational and research settings. The document provides several suggestions about data mining methodologies in the domains of depicted database management, data analysis, statistical learning, and machine learning. The training dataset accounts for 80% of the data, while the testing dataset accounts for the remaining 20%. The NB technique has the fastest runtime and lowest error rate (Table 3.1) [21].

TABLE 3.1 Comparison of the Naive Bayes, Decision tree, and k-Nearest Neighbors Categorization Algorithms' Accuracy and Time Complexity

Algorithm	Accuracy (%)	Time Taken (min)
Decision tree	52	713
K-nearest neighbors	46	1000
Naive bayes	53	608

3.7 CONCLUSION

Heart disease may progress to an unmanageable state. Cardiovascular diseases provide significant challenges in terms of treatment efficacy and contribute to a substantial annual mortality rate. Ignoring cardiac warning signs can be lethal. The high-stress levels of the general population, along with the prevalence of sedentary lifestyles, have only made the problem worse. Early detection of the disease enables effective control measures. The prioritization of exercise and good behavior should be upheld consistently. The use of tobacco and the consumption of an unhealthy diet are both factors that contribute to an elevated susceptibility to stroke and heart disease. It is recommended to consume five servings of fruits and vegetables daily. It is recommended that those with cardiovascular conditions restrict their consumption of salt to a maximum of one teaspoon per day. A significant drawback lies in the emphasis of these studies on the classification of cardiac disease rather than the processing of data for data mining. A dataset that has undergone the process of cleaning and trimming exhibits greater accuracy compared to a dataset containing missing values. The utilization of data cleaning and classification techniques enhances the precision of predictions. In the future, it is conceivable that an advanced intelligent system might potentially aid individuals diagnosed with heart disease in making informed decisions regarding optimal treatment options. The utilization of predictive models for heart disease has experienced a surge in popularity. After a heart disease diagnosis, patients have numerous therapy options. The utilization of data mining techniques on appropriate datasets can facilitate the identification of optimal treatment strategies.

KEYWORDS

- **arteries**
- **cardiovascular disease**
- **coronary heart disease**
- **decision tree**
- **electrocardiogram**
- **electronic health record**
- **fuzzy logic**

REFERENCES

1. LeCun, Y., Bengio, Y., & Hinton, G. (2015). Deep learning. *Nature*, *521*(7553), 436–444.
2. Johnson, K. W., Torres Soto, J., Glicksberg, B. S., Shameer, K., Miotto, R., Ali, M., Ashley, E., & Dudley, J. T. (2018). Artificial intelligence in cardiology. *Journal of the American College of Cardiology*, *71*(23), 2668–2679.
3. Phan, T. T., Abozguia, K., Nallur Shivu, G., Mahadevan, G., Ahmed, I., Williams, L., Dwivedi, G., Patel, K., Steendijk, P., Ashrafian, H., & Henning, A. (2009). Heart failure with preserved ejection fraction is characterized by dynamic impairment of active relaxation and contraction of the left ventricle on exercise and associated with myocardial energy deficiency. *Journal of the American College of Cardiology*, *54*(5), 402–409.
4. Souza Filho, E. M., Fernandes, F. D., Soares, C. L., Seixas, F. L., Santos, A. A., Gismondi, R. A., Mesquita, E. T., & Mesquita, C. T. (2020). Artificial intelligence in cardiology: Concepts, tools, and challenges—"The horse is the one who runs, you must be the jockey." *Arquivos Brasileiros de Cardiologia*, *114*, 718–725.

5. Kuo, F. C., Mar, B. G., Lindsley, R. C., & Lindeman, N. I. (2017). The relative utilities of genome-wide, gene panel, and individual gene sequencing in clinical practice. *Blood, The Journal of the American Society of Hematology*, *130*(4), 433–439.
6. Muse, E. D., Barrett, P. M., Steinhubl, S. R., & Topol, E. J. (2017). Towards a smart medical home. *The Lancet*, *389*(10067), 358.
7. Steinhubl, S. R., Muse, E. D., & Topol, E. J. (2015). The emerging field of mobile health. *Science Translational Medicine*, *7*(283), 283rv3.
8. Shameer, K., Badgeley, M. A., Miotto, R., Glicksberg, B. S., Morgan, J. W., & Dudley, J. T. (2017). Translational bioinformatics in the era of real-time biomedical, health care, and wellness data streams. *Briefings in Bioinformatics*, *18*(1), 105–124.
9. Visco, V., Izzo, C., Mancusi, C., Rispoli, A., Tedeschi, M., Virtuoso, N., Giano, A., Gioia, R., Melfi, A., Serio, B., & Rusciano, M. R. (2023). Artificial intelligence in hypertension management: An ace up your sleeve. *Journal of Cardiovascular Development and Disease*, *10*(2), 74.
10. Krittanawong, C., Zhang, H., Wang, Z., Aydar, M., & Kitai, T. (2017). Artificial intelligence in precision cardiovascular medicine. *Journal of the American College of Cardiology*, *69*(21), 2657–2664.
11. Abdolmanafi, A., Duong, L., Dahdah, N., & Cheriet, F. (2017). Deep feature learning for automatic tissue classification of coronary artery using optical coherence tomography. *Biomedical Optics Express*, *8*(2), 1203–1220.
12. Pouriyeh, S., Vahid, S., Arabnia, H. R., & Sannino, G. (2017). A comprehensive investigation on comparison of machine learning techniques on heart disease domain. *Proceedings of the 2017 International Conference on Machine Learning*.
13. Chauhan, R., Bajaj, P., Choudhary, K., & Gigras, Y. (2015). Framework to predict health diseases using attribute selection mechanism. In *2015 2nd International Conference on Computing for Sustainable Global Development (INDIACom)* (pp. 1880–1884). IEEE.
14. Chauhan, S., Rühaak, W., Khan, F., Enzmann, F., Mielke, P., Kersten, M., & Sass, I. (2016). Processing of rock core microtomography images: Using seven different machine learning algorithms. *Computers & Geosciences*, *86*, 120–128.
15. Chaurasia, V., & Pal, S. (2014). Data mining approach to detect heart diseases. *International Journal of Advanced Computer Science and Information Technology (IJACSIT)*, *2*, 56–66.
16. Fatima, M., & Pasha, M. (2017). Survey of machine learning algorithms for disease diagnostic. *Journal of Intelligent Learning Systems and Applications*, *9*(1), 1–14.
17. Lloyd-Jones, D. M., Hong, Y., Labarthe, D., Mozaffarian, D., Appel, L. J., Van Horn, L., Greenlund, K., Daniels, S., Nichol, G., Tomaselli, G. F., & Arnett, D. K. (2010). Defining and setting national goals for cardiovascular health promotion and disease reduction: The American Heart Association's strategic Impact Goal through 2020 and beyond. *Circulation*, *121*(4), 586–613.
18. Gudadhe, S. S., Thakare, A. D., & Oliva, D. (2023). Classification of intracranial hemorrhage CT images based on texture analysis using ensemble-based machine learning algorithms: A comparative study. *Biomedical Signal Processing and Control*, *84*, 104832.

19. Ordonez, C. (2006). Association rule discovery with the train and test approach for heart disease prediction. *IEEE Transactions on Information Technology in Biomedicine*, *10*(2), 334–343.
20. Bhatt, C. M., Patel, P., Ghetia, T., & Mazzeo, P. L. (2023). Effective heart disease prediction using machine learning techniques. *Algorithms*, *16*(2), 88.
21. Seh, A. H., & Chaurasia, P. K. (2019). A review on heart disease prediction using machine learning techniques. *International Journal of Management, IT and Engineering*, *9*(4), 208–224.

CHAPTER 4

Monitoring the Health Status of Thyroid Patients Using Machine Learning

ABSTRACT

Thyroid disease exhibits a high prevalence within the population. A precise and prompt diagnosis of this disease is of utmost importance. Although machine learning approaches can be used to diagnose this disease, it is important to note that the most reliable and widely accepted method is still a comprehensive set of laboratory tests and imaging examinations. Feature extraction with correlation is the most successful machine learning strategy for the classification of two different types of thyroid disease (hyperthyroidism and hypothyroidism). Support vector machines (SVMs), random forests (RFs), K-nearest neighbors (KNNs), decision trees (DTs), artificial neural networks (ANNs), and logistic regression (LR) are commonly employed as predictive models.

4.1 INTRODUCTION

Thyroid disease is a frequently neglected endocrine disease [1, 2]. According to the World Health Organization (WHO), thyroid gland diseases, which rank second in terms of prevalence after diabetes, represent the most common endocrine disorder. Approximately 1% of the population is affected by hypothyroidism, whereas hyperthyroidism is prevalent in approximately 2% of individuals. The male population constitutes approximately 10% of the total population, whereas the female population accounts for 90%. Dysfunctions of the brain, pituitary gland, or thyroid can all play a role in the development of hyperthyroidism or

Machine Learning in Healthcare: Advances and Future Prospects. Rishabha Malviya, Niranjan Kaushik, Tamanna Rai, M. P. Saraswathy, and Rajendra Awasthi (Authors)

hypothyroidism. There is a correlation between insufficient dietary iodine intake and a higher prevalence of goiter and active thyroid nodules in certain regions, with a reported incidence rate of 15%. The thyroid gland is susceptible not only to the development of malignant tumors but also to the detrimental impact of autoantibodies produced by the body [3]. It is well acknowledged among medical professionals that the timely identification, diagnosis, and treatment of diseases play a pivotal role in impeding the advancement of illnesses and preserving human lives. The early detection and differential diagnosis of several abnormalities have been shown to enhance treatment success rates [4].

The thyroid gland, referred to as the "butterfly gland," is a small gland located in the front region of the neck, exhibiting a resemblance to the shape of a butterfly. The brain synthesizes two biologically active thyroid hormones, namely levothyroxine (T4) and triiodothyronine (T3), which play a vital role in regulating thermoregulation, blood pressure, and cardiac rhythm, among various other physiological processes. Thyroid hormones are released by the thyroid, an endocrine gland. The thyroid hormones are transported throughout the body via the circulatory system. Thyroid hormones have a crucial role in facilitating the process of digestion, regulating the equilibrium of fluids and electrolytes, and performing various other functions within the body. Several hormones that are released by the thyroid gland are used as markers for assessing thyroid functionality. T3, T4, and TSH exemplify thyroid-stimulating hormones (TSHs). Hypothyroidism and hyperthyroidism are the prevailing forms of thyroid disease. The process of identifying patterns in large databases can be semi-automated through the utilization of data mining techniques [5].

Machine learning algorithms have been recognized as a very effective technique for addressing a diverse range of complex problems [6]. To enhance understanding regarding the role of machine learning algorithms in the categorization of thyroid disease, a study was conducted and a classification framework was developed. Classification, a data analysis technique, can be employed to diagnose and predict thyroid disease using machine learning algorithms [7]. The utilization of machine learning and artificial intelligence (AI) in the field of medicine can be traced back to its early stages [8]. There is a growing consensus that underscores the significance of healthcare solutions based on machine learning. Due to this rationale, it is anticipated by experts that machine learning will become pervasive in the domain of medicine [9].

4.2 THYROID FUNCTIONING

The pituitary gland exhibits diminished production of TSH because of elevated T4 levels, thereby leading to a deceleration of thyroid activity. Thyroxine (T4), which consists of four iodine molecules, is the primary thyroid hormone that the thyroid gland produces. The conversion of T4 to T3 (triiodothyronine) occurs through the release of an iodine particle, enabling T3 to exert its biological effects. The synthesis of T3 occurs mostly in organs such as the liver and brain. The release of the thyroid hormone T4 is regulated by the TSH. The TSH hormone is produced by the pituitary gland located in the brain. The release of TSH by the pituitary gland is dependent upon its detection of thyroxine (T4). The synthesis of TSH is enhanced in response to the detection of low levels of thyroxine (T4) by the pituitary gland. The synthesis of TSH by the pituitary gland stops after a particular amount of thyroxine (T4) has been produced. The thyroid and pituitary glands act as radiators and internal regulators. When the radiator is off, the indoor thermostat turns on the heater if the temperature falls below a certain threshold. The thermostat turns off the radiator when the temperature reaches an appropriate level. The thyroid and pituitary glands function in a manner analogous to home thermostats and light switches [10].

4.3 ARCHITECTURE OF THYROID PREDICTION SYSTEM

Machine learning is a branch of AI that is gradually permeating various academic disciplines. Algorithms provide the capability to acquire knowledge from previous errors through automatic learning techniques, which occur in a hidden manner [11]. The proliferation of machine learning can be attributed to the escalation in computational capabilities and the expanding volume of data accessible for processing. Classical epidemiology represents an innovative integration of contemporary data science with traditional epidemiology, enabling the utilization of computer-generated data to its fullest potential. These tools examine the intimate correlation between input and outcome, which holds significant clinical relevance. This enables the examination of extensive data sets [12]. One may easily be deceived by objective assessments of surgical observations, leading to modifications in surgical agreements. When attempting to resolve a surgical disagreement, it is crucial to elucidate the role played by the patient's acquaintance or relative in assisting with the procedure. Machine learning enables computers to use historical data to make accurate predictions about future events. The sensible component of the algorithm demonstrates a high level of accuracy in predicting outcomes, effectively replicating the intricate patterns observed in extensive and intricate datasets. Furthermore, it successfully captures the essence and characteristics of reliable data sources [13].

Machine learning can be effectively employed in the diagnosis and treatment of thyroid disease due to its multifactorial etiology and many treatment modalities [14]. This demonstrates the enormous potential of machine learning models and bolsters the growing trend towards precision medicine, in which each patient's care is meticulously customized. It is feasible to purposely build a large gap between supervised and unsupervised learning in the field of AI. Supervised learning methods develop a model that can predict new data that hasn't been seen before using "labeled" training data [15]. Given that unsupervised learning exclusively operates with unannotated data and depends on analogies and heuristics, it can be effectively employed to analyze a substantial corpus of unannotated genomics data. These techniques can be employed to generate labels intended for training a supervised model. This approach is valuable in the analysis and comprehension of intricate data that poses challenges in terms of human measurement [16]. Figure 4.1 illustrates a thyroid prediction system based on machine learning.

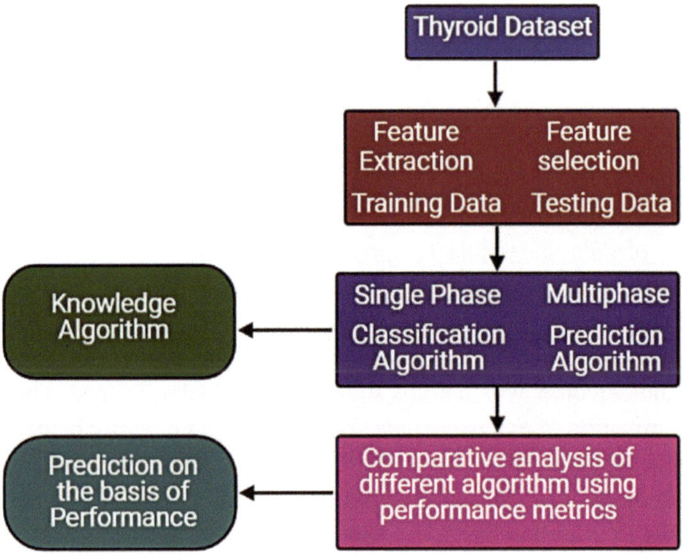

FIGURE 4.1 Illustration of a machine learning-based thyroid prediction system.

The conventional programming approach requires the systematic arrangement of data to achieve an anticipated result based on the provided input. Machine learning algorithms derive rules from labeled training data by analyzing the relationship between input and output variables. Machine learning has been demonstrated to be a highly efficient approach for analyzing enormous amounts of data, generating hidden information within databases, and adjusting to dynamic surroundings [17–20]. To minimize the difference between intended and attained outcomes, learning algorithms employ a technique known as feature weighting, wherein input variables (referred to as features) are assigned priority and significance based on their relevance in providing pertinent information. Machine learning enables the training of systems using extensive databases, wherein established machine learning techniques are employed to generate abstraction mechanisms or construct models. Subsequently, these mechanisms or models can be utilized to make predictions regarding upcoming events while ensuring the confidentiality of claimed predictions [18–21].

4.4 APPROACHES FOR THYROID PREDICTION

The objective of the proposed methodology is to ascertain prospective approaches for the treatment of thyroid disease by analyzing historical and contemporary patient records.

4.4.1 DATA COLLECTION

The initial dataset comprises individual-specific information encompassing age, date of birth, gender, medical conditions, occupation, educational attainment, marital status, biological sex, physical attributes such as height, weight, and body mass index, as well as clinical data about potential pregnancies and menstrual cycles in women. This clinical data encompasses observations related to the skin, heart, neck, abdomen, extremities, thorax, and eyes. The second source pertains to the medical record of the patient [22]. Each patient's file contains laboratory test results and notes from doctor's appointments. The process of patient identification involves the integration of two distinct data sources, resulting in the creation of a comprehensive and extensive dataset.

4.4.2 THE PROPOSED FEATURE MODEL

The features are obtained from an initial dataset that encompasses comprehensive information on the patients from multiple perspectives. Characteristics such as personal information, medical history (both immediate and extended), current health condition, blood test results, and hormone and thyroid levels might be cited as illustrative instances. The first feature set is refined to a limited number of attributes that primarily focus on patient data and variables related to thyroid function. This determination is made by an expert utilizing established criteria for evaluating the effectiveness of medical treatment [23]. Certain features are excluded from the analysis due to their presence in just a sample of patients, resulting in their absence in over half of the entire dataset. The specific values encompassed in this context are as follows: "increased" denotes instances where there is a necessity to augment the patient's dosage; "decreased" signifies situations where a reduction in the patient's dosage is warranted; "stable" indicates circumstances where the treatment regimen should remain unaltered; and "others" encompasses scenarios where it is imperative to suspend the treatment entirely for the patient.

4.4.3 CLASSIFIERS

Several machine learning classifiers, such as AdaBoost, gradient boosting, XGBC, and CatBoost algorithms, have been used to predict the therapy plan for patients with thyroid conditions. Based on the patient's medical history and current clinical status, the classifier provides a recommendation to the endocrinologist regarding the appropriate adjustment of the patient's LT4 dosage, which may involve an increase, decrease, or maintenance of the current dosage. To identify which algorithms best categorize each item within the dataset, distinct attributes are contrasted and analyzed. The boosting algorithms are a subset of the algorithms that have been chosen [24].

4.5 ALGORITHMS FOR MACHINE LEARNING

4.5.1 ARTIFICIAL NEURAL NETWORKS (ANNS)

Artificial neural networks (ANNs) receive influence from the structure and functionality of the human nervous system. They possess the ability to learn and simulate various types of functions, including those that use real-valued, discrete-valued, and vector-valued inputs, by employing a substantial number of interconnected units known as neurons. Backpropagation is widely regarded as the preferred method for learning in ANNs. The algorithms of the neural networks exhibit a tripartite structure. The design of the system comprises three layers, namely the input layer, the hidden layer, and the output layer. The input layer, located at the topmost level of the hierarchical structure, receives data from the outer layers. The hidden layer, positioned in the center, processes the received data. Lastly, the output layer, serving as the ultimate layer, disseminates the network's prediction. By utilizing this compact network, it is possible to classify the newly acquired data.

4.5.2 DECISION TREE (DT)

The decision tree (DT) classifier uses a graphical representation that bears resemblance to a tree structure. Within the context of DTs, it is crucial to note the existence of three different types of nodes, namely internal nodes, leaf nodes, and root nodes. An internal node of a tree symbolizes a test conducted on a specific property, while a leaf node signifies the distribution of a particular class. Lastly, the root node of a tree indicates the apex of the tree. C4.5 and ID3 are the two primary algorithms employed in a DT-based model for the diagnosis and prediction of thyroid diseases, offering a comprehensive approach. DTs are frequently employed in the medical domain, particularly in the context of diagnosing thyroid problems [25].

4.5.3 K-NEAREST NEIGHBOR (KNN)

When a training tuple is sent to the KNN algorithm, it is stored for later use when a test tuple is provided for evaluation. The term "lazy learner" is used to describe a machine learning algorithm that retains its training data, or "instances," for future reference [26, 27]. The determination of the number of neighbors considered for categorization is contingent upon the selection of a positive integer, denoted as "k." In the realm of distance measures, the concept of "closeness" is typically characterized by metrics such as "Euclidean distance" and "Manhattan distance."

4.5.4 SUPPORT VECTOR MACHINE (SVM)

Support vector machines (SVMs) enable precise analysis through the utilization of a wide range of research methodologies. The SVM is a computational method that utilizes a hyperplane separation algorithm to facilitate the analysis of data sample distribution [28]. The SVM classifier can generate one or multiple hyperplanes inside a high-dimensional domain. The training data can be divided into positive and negative sets by employing a hyperplane.

The machine learning library at the University of California, Irvine, has been used to obtain the datasets associated with thyroid disease [29]. The endeavor can be divided into two primary phases. The first stage involved choosing a subset of the thyroid datasets using a customized approach based on mutual information and ANN prediction [30]. Neural networks have demonstrated successful implementation in specific sectors of the medical field, particularly in the realm of disease diagnosis and interpretation. The selection of a particular feature selection technique is dependent upon the level of reliability exhibited by the study conducted on datasets encompassing data about thyroid disorders [31].

4.7 CONCLUSION

In future studies, we will investigate whether tailored machine learning can detect thyroid disorders. A few simple investigations have helped diagnose thyroid disease in recent years. Scientific research has shown that neural networks outperform other methods. However, this is to be expected given the effectiveness of both the SVM and the DT. Researchers have made significant progress toward precisely identifying thyroid problems. A patient with more characteristics requires more time and money for clinical evaluation. Thus, it is critical to develop algorithms and thyroid illness predictive models that can diagnose the condition with minimal patient input, thereby saving both time and money.

KEYWORDS

- **decision tree**
- **K-nearest neighbors**
- **logistic regression**
- **machine learning**
- **naïve bayes**
- **random forest**
- **support vector machines**
- **thyroid diseases**

REFERENCES

1. Erol, R., Oğulata, S. N., Şahin, C., & Alparslan, Z. N. (2008). A radial basis function neural network (RBFNN) approach for structural classification of thyroid diseases. *Journal of Medical Systems*, *32*, 215–220.
2. Keleş, A., & Keleş, A. (2008). ESTDD: Expert system for thyroid diseases diagnosis. *Expert Systems with Applications*, *34*(1), 242–246.
3. Cichosz, S. L., Johansen, M. D., & Hejlesen, O. (2016). Toward big data analytics: A review of predictive models in the management of diabetes and its complications. *Journal of Diabetes Science and Technology*, *10*(1), 27–34.
4. Kourou, K., Exarchos, T. P., Exarchos, K. P., Karamouzis, M. V., & Fotiadis, D. I. (2015). Machine learning applications in cancer prognosis and prediction. *Computational and Structural Biotechnology Journal*, *13*, 8–17.
5. Khushf, G. (1998). A radical rupture in the paradigm of modern medicine: Conflicts of interest, fiduciary obligations, and the scientific ideal. *The Journal of Medicine and Philosophy*, *23*(1), 98–122.
6. Aswad, S. A., & Sonuç, E. (2020). Classification of VPN network traffic flow using time-related features on Apache Spark. In *2020 4th International Symposium on Multidisciplinary Studies and Innovative Technologies (ISMSIT)* (pp. 1–8). IEEE.
7. Banu, G. R. (2016). A role of decision tree classification data mining technique in diagnosing thyroid disease. *International Journal of Computer Sciences and Engineering*, *4*(11), 64–70.
8. LeFevre, M. L., & US Preventive Services Task Force*. (2015). Screening for thyroid dysfunction: US Preventive Services Task Force recommendation statement. *Annals of Internal Medicine*, *162*(9), 641–650.
9. Murdoch, T. B., & Detsky, A. S. (2013). The inevitable application of big data to health care. *JAMA*, *309*(13), 1351–1352.

10. El-Omari, N. K., Al-Omari, A. H., Al-Ibrahim, A. M., & Alwada'n, T. (2017). *International Journal of Basic Sciences and Applied Computing (IJBSAC)*, *6*(4), 1–9.
11. Bouton, C. E., Shaikhouni, A., Annetta, N. V., Bockbrader, M. A., Friedenberg, D. A., Nielson, D. M., Sharma, G., Sederberg, P. B., Glenn, B. C., Mysiw, W. J., & Morgan, A. G. (2016). Restoring cortical control of functional movement in a human with quadriplegia. *Nature*, *533*(7602), 247–250.
12. Bennett, D., Silverstein, S. M., & Niv, Y. (2019). The two cultures of computational psychiatry. *JAMA Psychiatry*, *76*(6), 563–564.
13. Parthiban, G., & Srivatsa, S. K. (2012). Applying machine learning methods in diagnosing heart disease for diabetic patients. *International Journal of Applied Information Systems*, *3*(7), 25–30.
14. Verma, L., Srivastava, S., & Negi, P. C. (2016). A hybrid data mining model to predict coronary artery disease cases using non-invasive clinical data. *Journal of Medical Systems*, *40*, 1–7.
15. Ehrenstein, V., Nielsen, H., Pedersen, A. B., Johnsen, S. P., & Pedersen, L. (2017). Clinical epidemiology in the era of big data: New opportunities, familiar challenges. *Clinical Epidemiology*, 245–250.
16. Ghahramani, Z. (2015). Probabilistic machine learning and artificial intelligence. *Nature*, *521*(7553), 452–459.
17. Acosta, F. L., McClendon, J., O'Shaughnessy, B. A., Koller, H., Neal, C. J., Meier, O., Ames, C. P., Koski, T. R., & Ondra, S. L. (2011). Morbidity and mortality after spinal deformity surgery in patients 75 years and older: Complications and predictive factors. *Journal of Neurosurgery: Spine*, *15*(6), 667–674.
18. Deo, R. C. (2015). Machine learning in medicine. *Circulation*, *132*(20), 1920–1930.
19. Austin, P. C., Tu, J. V., Ho, J. E., Levy, D., & Lee, D. S. (2013). Using methods from the data-mining and machine-learning literature for disease classification and prediction: A case study examining classification of heart failure subtypes. *Journal of Clinical Epidemiology*, *66*(4), 398–407.
20. Pandey, A. K., Pandey, P., Jaiswal, K. L., & Sen, A. K. (2013). A heart disease prediction model using decision tree. *IOSR Journal of Computer Engineering (IOSR-JCE)*, *12*(6), 83–86.
21. Kousiouris, G., Menychtas, A., Kyriazis, D., Gogouvitis, S., & Varvarigou, T. (2014). Dynamic, behavioral-based estimation of resource provisioning based on high-level application terms in cloud platforms. *Future Generation Computer Systems*, *32*, 27–40.
22. Fang, W., Lu, Z., Wu, J., & Cao, Z. (2012). Rpps: A novel resource prediction and provisioning scheme in cloud data center. In *2012 IEEE Ninth International Conference on Services Computing* (pp. 609–616). IEEE.
23. Chawla, N. V., Bowyer, K. W., Hall, L. O., & Kegelmeyer, W. P. (2002). SMOTE: Synthetic minority over-sampling technique. *Journal of Artificial Intelligence Research*, *16*, 321–357.
24. Freund, Y., & Schapire, R. E. (1997). A decision-theoretic generalization of on-line learning and an application to boosting. *Journal of Computer and System Sciences*, *55*(1), 119–139.

25. Chung, T. H. (1994). Approximate methods for sequential decision making using expert advice. In *Proceedings of the Seventh Annual Conference on Computational Learning Theory* (pp. 183–189).
26. Lagani, V., Koumakis, L., Chiarugi, F., Lakasing, E., & Tsamardinos, I. (2013). A systematic review of predictive risk models for diabetes complications based on large scale clinical studies. *Journal of Diabetes and Its Complications*, *27*(4), 407–413.
27. Ahmed, S. S., Dey, N., Ashour, A. S., Sifaki-Pistolla, D., Bălas-Timar, D., Balas, V. E., & Tavares, J. M. (2017). Effect of fuzzy partitioning in Crohn's disease classification: A neuro-fuzzy-based approach. *Medical & Biological Engineering & Computing*, *55*, 101–115.
28. Pisner, D. A., & Schnyer, D. M. (2020). Support vector machine. In *Machine Learning* (pp. 101–121). Academic Press.
29. Kwon, M. R., Shin, J. H., Park, H., Cho, H., Hahn, S. Y., & Park, K. W. (2020). Radiomics study of thyroid ultrasound for predicting BRAF mutation in papillary thyroid carcinoma: Preliminary results. *American Journal of Neuroradiology*, *41*(4), 700–705.
30. Zhang, M., Yang, Y., Zhang, H., Shen, F., & Zhang, D. (2016). L2, p-norm and sample constraint-based feature selection and classification for AD diagnosis. *Neurocomputing*, *195*, 104–111.
31. Bonci, A., Fiori, S., Higashi, H., Tanaka, T., & Verdini, F. (2021). An introductory tutorial on brain–computer interfaces and their applications. *Electronics*, *10*(5), 560.

CHAPTER 5

Machine Learning-Based Wearable Devices for Healthcare Applications

ABSTRACT

There has been a notable surge in the inclination towards the utilization of machine learning and artificial intelligence (AI) within the discipline of healthcare in the past few years. Due to its application in the surveillance of wearable technology, which tracks human behavior and physiological information, as well as its role in aiding disease identification, this technology holds considerable promise for implementation in senior care, patient monitoring, and therapy. The significance of wearable healthcare devices has increased due to advancements in medical sensors and the downsizing of electronic circuits. This chapter presents an overview of the key topics in modern machine learning related to these devices.

5.1 INTRODUCTION

The term "wearable technology" comprises an extensive range of electronic devices that have been specifically intended to be worn either directly on the body or near it. The latter categorization often includes mobile phones, which have increasingly played a crucial role in enabling the advancement of wearable technology [1]. The classification of smartphones as wearable technologies has been a topic of debate. However, the widespread use of smartphones has led to an upsurge in interest in wearable technologies as effective tools for improving quality of life [2]. The primary factor contributing to this phenomenon may be attributed to

Machine Learning in Healthcare: Advances and Future Prospects. Rishabha Malviya, Niranjan Kaushik, Tamanna Rai, M. P. Saraswathy, and Rajendra Awasthi (Authors)

the widespread availability of supplementary software applications. While this has fostered an environment of innovation, it has also impeded the development of cohesive application design. Consequently, users are faced with an overwhelming abundance of options. The present technological advancements in smartphones have made them more portable and capable of performing complex computations. This has significant implications for the field of bioassays, as it allows for the quick and dependable execution of these tests in any location and at any time [3]. In essence, wearable technology may be categorized into two distinct types: primary and secondary. The primary type functions independently and acts as a central hub, facilitating the connection between different electrical and data systems. On the other hand, the secondary type is designed to capture events or collect measurements, which are then transmitted to the main wearable device [4]. The inclusion of smart textiles, which include the ability to measure or respond to input from the user or the environment based on their physical properties, can also be considered within this category [2]. The practice of integrating electronics or unconventional tailoring materials into clothes or directly onto the human body is currently limited to enthusiasts with a futuristic vision. However, there are indications that this situation may transform the near future.

Accelerometers, optical sensors, temperature sensors, and biometric sensors are typical examples of the various types of sensors that could be included in a wearable device to continuously monitor a diverse array of human signals. Despite the potential lack of accuracy in the readings from these sensors, they can still be used in circumstances where more permanent medical equipment is not readily available, subject to their particular application [1, 2]. Algorithms employing machine learning techniques can detect and recognize meaningful patterns within the data produced by sensors in Internet of Medical Things devices, as well as from human engagement with those devices. The technology mentioned herein exhibits significant potential in the realm of health applications, specifically in the areas of vital sign monitoring, disease detection, recognizing falls, and stress identification. The application of machine learning techniques for the analysis of data obtained from wearable sensors connected to human subjects has been the focus of extensive study in the past 10 years. Despite extensive research and the remarkable proliferation of wearable devices, particularly smartwatches, the commercialization of machine learning applications remains limited in scope. One instance of such functionality is represented by alarms that provide notifications regarding irregular heartbeats [3]. In 2018, the Food and Drug Administration (FDA) granted approval to the Apple Watch, accompanied by an extensive compilation of potential dangers and cautionary statements.

This chapter provides an overview of the existing research on the integration of machine learning techniques into wearable devices. Concerns about the development, storage, energy consumption, user acceptability, dependability, information exchange, confidentiality, and the confidentiality of wearable machine learning applications are addressed. This chapter delves into the approaches employed for privacy-preserving machine learning training and inference (Figure 5.1).

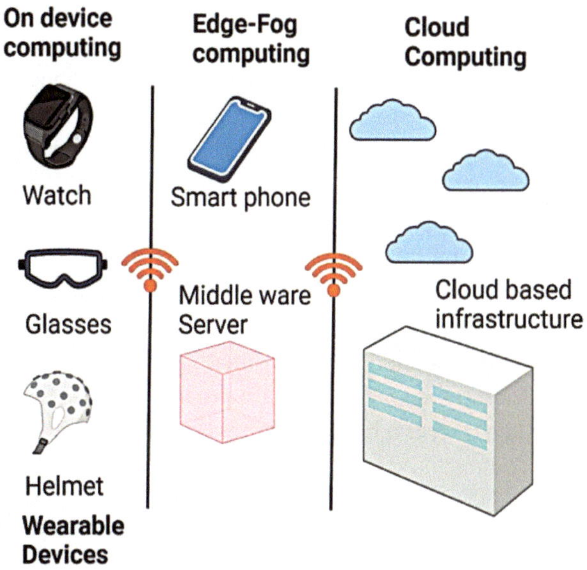

FIGURE 5.1 Illustration of the workflow of the wearable device application model.

5.2 SIGNALS OF WEARABLE DEVICES USED IN LEARNING

A wearable device can be worn that contains sensors capable of gathering data about the human body in a non-invasive manner. Several well-known signals and markers on the human body can be read to determine vital signs and other information about the patient's health and psychological status. Skin temperature sensors can be utilized as an illustrative instance to demonstrate the application of electrodermal activity sensors, also referred to as galvanic skin response sensors. These sensors are capable of capturing variations in skin conductance that are linked to the activity of the sympathetic nervous system [2, 5, 6]. Another example involves the use of an electrocardiogram (ECG) sensor to capture and measure the electrical fluctuations in the dermal layer of the body resulting from cardiac contractions [7–10]. Electromyography (EMG) and electroencephalogram (EEG) sensors are utilized to record electrical activity in the brain and muscle and nerve cell health, respectively [11–17]. With the help of an optical photoplethysmography sensor, a patient's pulse rate and heart rate variability can be determined by measuring their blood volume pulse, as explained in Refs. [1, 18–20]. A photoplethysmography sensor can enable an approximate estimation of blood oxygen saturation levels (SpO_2) [20–22]. Applications that make use of the data collected by accelerometers, gyroscopes, and magnetometers concerning a user's health and activity level are becoming increasingly popular [1, 23–37]. An electrooculogram is created by placing electrodes on the eye, and an electrogastrogram is created by placing electrodes on the stomach [38, 39]. Galvanic skin response and ECG signal measurements show that humans have varying autonomic responses to different stimuli [40]. This is a significant challenge for models attempting to include such sensory inputs. The inputs have a wide range of potential applications, including the use of aromatherapy and nutrition therapy in the treatment of neuropsychiatric and eating disorders. In certain instances, it is imperative to maintain vigilance over the broader context.

5.3 MACHINE LEARNING FOR WEARABLE DEVICES

Machine learning enables wearable devices to collect information through their experiences and autonomously make decisions or judgments without being specifically programmed for specific conditions. The classification of machine learning as supervised, semi-supervised, unsupervised, or reinforced depends on the characteristics of the instructional data that is provided. The documentation of historical events serves as a valuable resource for acquiring knowledge through the analysis of labeled or unlabeled data. In labeled data, the dependent variable can be quantitative or categorical. Machine learning serves several objectives, encompassing classification, regression, and clustering. Classification involves predicting category output variables; regression deals with numerical labels; and clustering addresses the analysis of unlabeled data. Most of the research conducted on machine learning in wearable technology focuses on the classification of data [41, 42]. A smaller portion of studies explores the potential for grouping data, and only a limited number of studies can be considered regression issues [43]. Over the past 10 years, there has been a surge in research on the application of machine learning techniques to physical signals, with applications in fitness tracking, senior care, and health monitoring. The potential applications of activity recognition encompass various aspects of health monitoring, including predicting and assessing sleep quality, detecting falls, monitoring epilepsy and diabetes, and predicting and monitoring vital signs [44]. Additionally, activity recognition can be utilized for monitoring human activities related to health and fitness, among other functions. Numerous disciplines have explored the possibility of wearable technology, including tachycardia monitoring, stress monitoring, and rehabilitative services.

5.3.1 ROLE OF WEARABLE DEVICES IN SLEEP HEALTH

To investigate the relationship between physical activity and sleep, researchers have employed a variety of devices, including actigraphy sensors, standardized questionnaires, and polysomnography. Each of these methods has significance in clinical practice, especially in diagnosing sleep disorders [45]. Polysomnography is commonly recognized as the preferred method for identifying sleep-related respiratory issues due to its extensive monitoring of the patient's physiological conditions for the whole duration of the night. Overnight, these sensors keep a patient under close observation [46]. This can be used to analyze different types of sleep disorders. Polysomnography is normally done overnight due to its complexity. Although a polysomnography examination can be done in the patient's home, it is challenging to carry out with portable solutions [47, 48]. There is a need for novel methodologies to explore the correlation between the waking and sleeping habits of an individual. Actigraphy is a technique that was developed in the early 1990s to investigate the patterns and characteristics of sleep using portable electronic devices [49]. Actigraphy can monitor not just how active a person is throughout the day but also how well they sleep at night. Actigraphy has become popular rapidly because it is more reliable than subjective sleep diaries and behavior logs [50].

5.3.2 ROLE OF WEARABLE DEVICES IN SEIZURE DETECTION

Machine learning techniques are employed to analyze EEG datasets obtained from individuals diagnosed with epilepsy. Genetic algorithms that can recognize epileptic activity in EEG data have been extensively investigated. The Kaggle competition employs implanted EEG records obtained from both humans and canines to develop a comprehensive system for detecting seizures on a global scale [51]. The approach used for identifying seizures from long-term EEG involved the utilization of a random forest (RF) classifier, which exhibited a high level of performance with an area under the curve above 0.97. Machine learning algorithms can identify seizures through the analysis of scalp electroencephalography data. These algorithms have false-positive rates ranging from 0.1 to 5 per hour, along with sensitivities ranging from 75% to 90% [52]. Both seizures and interictal epileptiform discharges hold similar significance in the diagnosis of epilepsy. The utilization of video EEG monitoring for extended durations within residential settings is gaining popularity. The use of automated and dependable spike detection techniques greatly facilitates the processing of large-scale datasets. Machine learning methods removed spike-free periods from electroencephalography [53]. Patients with generalized epilepsy had their epileptic discharges clinically assessed and quantified using deep learning (DL) [54, 55].

5.3.3 MACHINE LEARNING-BASED SEIZURE PREDICTION USING WEARABLE DEVICES

Various machine-learning approaches have been used to detect and predict seizure events based on WD signals recorded throughout phase 0 to phase 2 trials [56]. Tonic-clonic seizures can be detected by the utilization of support vector machine (SVM) models, which are trained on data obtained from recordings of accelerometry and electrodermal activity [57, 58]. These signals have also been used in an algorithm that combines k-nearest neighbor (KNN) and RF features [59]. The FDA-approved wrist-worn seizure-detection watch was used in all the subsequent investigations.

Nevertheless, the assessment of the comparative efficacy of various machine learning algorithms is a complex task due to the lack of recorded seizures and the absence of consensus among scientists over the precise meaning of seizures. Previous research has exclusively documented the retrospective efficacy of smartwatches in seizure detection. As far as current information is concerned, there is a lack of published findings from prospective studies utilizing the same technology. EMG signals have commonly been employed as a feature in machine learning systems for seizure detection when combined with data obtained from sensors worn on the wrist. Larsen et al. successfully attained a high level of sensitivity (median = 1.0, min = 0.5) using surface EMG data obtained from deltoid electrodes. These data were employed to extract relevant features to train a RF classifier to detect generalized tonic-clonic seizures (GTCSs) [60]. In addition to seizures, machine learning techniques have also been employed to detect pre-ictal signal features from wearable devices. Heart rate variability has been extensively used for seizure prediction [61], and preictal changes in heart rate have been documented. Seizure prediction using ECG has recently been made possible with the application of DL algorithms [62]. Another study developed an algorithm that could predict seizures for individuals based on ECG data and an SVM. This study demonstrated an average sensitivity of 89% in 15 people with varied seizures, with predictive signals obtained up to 20 minutes before episodes [63]. Although there have been initial indications of potential in seizure-forecasting research, the evaluation of heart rate and wearable devices in a prospective setting has not been conducted yet [63, 64]. Comparing machine learning algorithms across different devices, seizure type definitions, and inclusion criteria remains a persistent challenge. The provision of extensive, standardized datasets containing signals from individuals with epilepsy, like the approach taken with EEG recordings, would serve to address the challenges that have been faced so far.

5.3.4 ROLE OF WEARABLE IN STRESS DETECTION

The stress detection survey incorporated various physiological and behavioral indicators, including heart rate, blood volume pulse, interbeat intervals (RR intervals), electrodermal activity, temperature, and behavioral aspects [18]. Electrodermal activity and heart rate are the best stress indicators.

The issue of the remote monitoring of children's stress levels to safeguard their well-being has been widely investigated [65, 66]. The investigation of utilizing a neural network to identify metabolic syndrome symptoms in children with autism spectrum disorder has been conducted due to the potential exacerbation of these symptoms by stress [22]. Biosignals, eye monitors, microphones, cameras, and mobile interactions have been used to study mental illnesses like anxiety, bipolar disorder, and depression [67, 68].

5.3.5 ROLE OF WEARABLES IN HYDRATION MONITORING

Various individuals, including athletes, military personnel deployed in combat zones, those working in high-temperature situations, and elderly individuals who may experience difficulties in expressing their thirst, all have a vested interest in addressing the issue of monitoring hydration levels. Currently, biochemical sensors are being utilized to monitor the electrolyte concentration in sweat and, consequently, the hydration level of an individual [69]. However, there are simultaneous efforts in the field of machine learning aimed at acquiring knowledge from diverse biological signals to detect dehydration. For instance, one such effort focuses on determining how dehydration-induced cognitive stress affects the body's autonomic responses [70]. To identify dehydration, ECG signal characteristics that were annotated with resting heart rate variability, heart rate variability during exercise, and heart rate variability following rehydration have been employed [71]. Electrodermal activity and heart rate variability characteristics from photoplethysmography data can be used to successfully find moderately dehydrated people. The researchers utilized data from many sensors to estimate the user's most recent consumption of beverages, thereby facilitating the collection of data and enabling modifications within the application.

5.3.6 ROLE OF WEARABLES IN DIABETES MONITORING

Wearables have been developed to monitor the blood glucose level, body temperature, and physical activity of individuals diagnosed with diabetes. This data is subsequently transmitted to a central base station through a smartphone that is connected to a 5G network. The device uses advanced AI and machine learning methodologies to effectively analyze the data, enabling users to exert enhanced management over their glucose levels and proactively predict any health concerns. The utilization of denser and smaller cells in 5G networks has the potential to significantly enhance the data transmission rate experienced by clients. There is a wide variety of devices to consider in the design of 5G networks. There are numerous types of sensors and equipment used in intelligent healthcare, and each provides a unique set of data that necessitates the use of 5G networks. For effective management of healthcare systems, data must be analyzed and used, which necessitates several network characteristics such as mobility, charging, security, policy administration, dependability, and latency. The implementation of 5G technology enables individuals to maintain communication with their healthcare providers, thereby facilitating the effective management of health conditions and potentially reducing healthcare costs. Without the necessity of a patient's physical presence, doctors can provide superior care anywhere in the world. Numerous applications in healthcare have the potential to use the substantial data transfer rates and constant reliability offered by 5G technology. The proposed system consists of four layers: layer one, which includes sensors; layer two, responsible for data collection; layer three, focused on transmission; and layer four, dedicated to the database.

5.3.6.1 LAYER 1: SENSORS

This layer contains the sensors that measure blood glucose, temperature, and movement. The ESP8266 module, which wirelessly connects the sensors and sends data to the patient's mobile device, is likewise located at this layer. Therefore, the sensors carry the responsibility for collecting and transmitting the data to the patient's smartphone.

5.3.6.2 LAYER 2: DATA ACQUISITION LAYER

The smartphone and data-gathering applications that belong to the patient are located in this layer. The mobile application displays the sensor readings. The 5G network enables several simultaneous connections within each coverage area, facilitating the delivery of data to the base station. The primary goal is to enhance the capacity to accommodate a tenfold increase in the number of devices within a given area of one square kilometer, surpassing the capabilities of the 4G network.

5.3.6.3 LAYER 3: TRANSMISSION LAYER

The data is sent over 5G from the patient's phone to the database and back to the doctor's phone for review.

5.3.6.4 LAYER 4: DATABASE LAYER

This location collects and stores data from numerous sensors before various AI algorithms process it. The server determines if the data is positive (true positive (TP)) or negative (false negative (FN)) using machine learning techniques. When an abnormality is found, the system notifies the user with an alert. A server-generated message is transmitted to the computer system operated by the doctor. Upon analyzing the notice, the doctor proceeds to transmit a text message to the patient with instructions about the prescribed treatment [71].

5.3.7 ROLE OF WEARABLES IN ARRHYTHMIA DETECTION

Consumer-grade wristbands and smartwatches often include major functionalities for measuring heart rate. There has been a notable surge in the number of commercial wearable devices that are specially engineered for detecting arrhythmia. The definition of a normal heart rate encompasses a range of 60 to 100 beats per minute (bpm). Atrial fibrillation, classified as a form of cardiac arrhythmia, is characterized by rapid and irregular contractions of the atrial chambers of the heart. Apple conducted a clinical study with a total of 4,19,297 participants, wherein photoplethysmography sensors integrated into Apple wristwatch patches were employed for the detection of atrial fibrillation. However, the method employed by the company did not use machine learning techniques; rather, it relied on a proprietary threshold that was created from data about the extent of dispersion observed in inter-peak intervals to assess irregularity. The study was undertaken by Apple and subsequently published in the Journal of the American College of Cardiology. Participants who exhibited indications of anomalies were asked to undergo ambulatory ECG monitoring with ECG patches during a specified period of observation and analysis. However, only 34% of the participants (450 individuals) agreed to this request. The introduction of the photoplethysmography signal was proposed as a potential method to address this issue [35]. Comparable results were obtained when monitoring oxygen saturation in the blood of individuals with atrial fibrillation using both classic pulse oximeters and the cardio tracker ring. According to the findings of the study, it was observed that all other variants of the SVM exhibited superior performance compared to a convolutional neural network (CNN) [72]. Even when the most unfavorable outcome for the 10-second recordings was considered, the level of accuracy remained at 94.7%. Although photoplethysmography signals have some problems, such as noise from motion artifacts, it is possible to consider using ring photoplethysmography-based wearables as an alternative to ECG-based methods for detecting atrial fibrillation. Even though photoplethysmography signals have their own set of constraints, considering apprehensions over the occurrence of false positives in cases of atrial tachyarrhythmia, it has been proposed that the examination of extended intervals be undertaken for photoplethysmography signals. The DL model achieved an accuracy rate of 89% when trained using data from both ECG and photoplethysmography sensors [73].

5.4 CHALLENGES OF WEARABLE TECHNOLOGY

5.4.1 DATA AVAILABILITY AND RELIABILITY

A substantial amount of data gathering is required to effectively train a machine learning model, particularly in the context of medical applications, to enable reliable prediction of future events based on historical data. To ascertain the reliability of data, it is imperative to conduct many clinical trials that employ various approaches and freely disclose their conclusions [74]. Additionally, it is crucial to uncover promising new pathways for further investigation. It is imperative to develop and regulate medico-legal considerations [75]. An illustrative framework reported by Nelson et al. offers a systematic approach to address the challenges associated with data reliability in the context of heart rate data power consumption. This framework encompasses the areas of research data collection and organization, data preprocessing and preparation, as well as reporting and analysis. It is worth noting that wearable technology, despite its potential benefits, is hindered by limitations such as high power consumption and limited battery life [76]. The acquisition of physiological data from wearable devices, in conjunction with the utilization of machine learning algorithms, necessitates a greater amount of electricity. The capabilities of even the most advanced commercial smartwatch are limited to monitoring basic activities such as walking and jogging, offering only approximate measurements of the wearer's heart rate and oxygen saturation, and exhibiting a maximum battery life of a few weeks. A minimal charging duration of a few hours may be sufficient for wearables designed to consistently monitor a diverse array of vital indicators to promptly notify users of any anomalous situations. The amount of data recorded on the device and transferred through the communication channel to the edge or cloud is contingent upon various factors, such as the type of board utilized, the quantity and nature of biosensors employed, the operating system and additional software operating on the board, the wearable display, the rate at which data is logged, and the rate at which data is transmitted. The use of electrical power is more pronounced in storage and processing activities compared to transmission and reception. The improvement in energy conservation encompasses various aspects, such as the integration of dedicated embedded circuits designed for machine learning algorithms [77, 78], the implementation of data reduction techniques [79–81], the use of data management strategies [83], the practice of data offloading [84, 85], and the development of self-powered wearable devices [86, 87].

5.4.2 MODEL SELECTION AND RELIABILITY

Researchers looked into cross-validation techniques for testing how well machine learning models work using data that wasn't known before. They found that using data from different sources to confirm results tends to make machine learning algorithm predictions more accurate and effective than they are. On the other hand, subject-wise cross-validation tends to lead to an underestimation of these metrics. These results were derived from a comparative analysis of the two methodologies. This provides support for the conclusions drawn in the study. However, several researchers have raised concerns regarding the applicability of record-wise cross-validation, arguing that it may not be suitable due to the absence of within-subject dependence among the observations. The investigators asserted that the feasibility of this outcome resulted from the lack of any discernible association between the recorded data and the observed phenomena. Various techniques, such as repeated test-train split, shuffle-split, repeated K-fold, Monte Carlo cross-validation, and the utilization of large fold numbers, have been employed to mitigate overfitting and ensure generalization. The utilization of models has proven to be advantageous for the development and implementation of wearable technology. The process of selecting a wearable device model encompasses a multitude of parameters. The optimization of the assessment measure has the potential to enhance the accuracy of classification or regression problems. Utilizing an ensemble of models to optimize outcomes is a common practice. The successful implementation of healthcare-related wearable device applications necessitates the utilization of user-friendly categorization, regression, and clustering findings. Tree-based models are often considered to be less complex in comparison to neural network-based models. The memory and model size of the wearable devices are additional factors to consider. The limited processing capabilities of wearable electronics have posed computational challenges for tasks such as inference and online training for customization until recent advancements. The technique of personalization involves the utilization of DL and transfer learning algorithms on the local device [88–90].

5.4.3 COMMUNICATION

In an Internet of Things edge architecture, a wearable and an EDM machine can communicate with one another via intra-device communication protocols such as radio-frequency identification, Bluetooth, Zigbee, near field communication, and ultra-wideband. Bluetooth is a popular choice for a wide range of applications because of its low power needs [91]. According to Bluetooth® Core Specification Version 5.0, a maximum of seven devices can be concurrently connected. However, in practical scenarios where a smartphone is connected to more than two devices, there is a noticeable decline in performance and the occurrence of pairing difficulties. Before deciding on a communication method, it is important to consider factors such as the maximum distance between the wearable and the edge device, the amount of data that needs to be sent from the wearable to the edge device, and the maximum amount of delay that can be tolerated [92]. The Internet Protocol and the Transport Control Protocol, or User Datagram

Protocol, at the network level, make it possible for edge devices to communicate with remote services or for wearables to communicate with remote services directly using the Internet. Transport Control Protocols or Internet Protocols are preferred for sending sensitive data over a wide-area network, such as medical records or AI model parameters. The hypertext transfer protocol (HTTP) is extensively used for requesting and responding to communication at the application layer between edge devices and cloud services. Although HTTP is known for its high resource requirements, it is often recommended to deploy it on edge or fog devices that possess ample processing capabilities and memory. Additionally, Transport Layer Security is frequently employed to secure HTTP communication via the transport control protocol. Lightweight application layer protocols include constrained application protocols, advanced message queuing protocols, and message queuing telemetry transfers [93]. Message queuing telemetry transfer has emerged as the prevailing standard for publish/subscribe models in the domains of the Internet of Things and wearable technology, mostly because of its minimal resource demands and extensive adoption. This technology facilitates bidirectional communication between a wearable device and an edge device, enabling the transmission of data from one device to multiple devices simultaneously. Both channels are susceptible to various network security vulnerabilities that are commonly encountered, mostly because of the protocols and network layers on which they are based.

5.4.4 SECURITY AND PRIVACY

The accelerometer and gyroscope of a smartwatch are capable of stealing credit card numbers and passwords. Denial-of-service and ransomware attacks against the Internet of Medical Things have the potential to severely interrupt medical services, leading to potentially fatal consequences. Machine learning services are employed to evaluate sensor data and personal data obtained from wearable devices integrated into a health or fitness tracking system to identify patterns and make prognostications. While some users might consider sharing this to be no problem, many end users are concerned about how their data will be utilized and protected. Global information security standards have been implemented to protect sensitive information and medical records. However, the emergence of wearables and other internet of medical things devices has introduced substantial risks to the security of this data [94].

5.5 CONCLUSION

The use of wearable technology has experienced a notable surge in popularity in recent years. Due to the extensive ongoing research on the utilization of AI solutions in healthcare-related professions, wearable devices have transitioned from being considered optional to becoming essential tools for remotely monitoring patients and detecting various physiological abnormalities. This chapter provides a comprehensive overview of the methodologies, instruments, and datasets employed in various studies about machine learning tasks that are relevant to healthcare wearable devices. We discuss potential solutions to challenges with using machine learning programs on wearable devices. Several elements need to be taken into consideration when evaluating deployment options, including the availability of resources such as power, memory, and storage. Additionally, it is important to assess the utility and user satisfaction of the chosen deployment strategy. Furthermore, the availability and dependability of data, as well as the effectiveness of communication, confidentiality, and secrecy measures, should also be considered. Data availability, dependability, and privacy are important issues that necessitate further investigation to facilitate the efficient and effective utilization of data derived from wearable devices for learning purposes.

KEYWORDS

- **actigraphy**
- **biosignals**
- **electromyography**
- **eye sensors**
- **galvanic skin response**
- **stress detection**
- **wearable devices**

REFERENCES

1. Thorp, E. O. (1998). The invention of the first wearable computer. In *Digest of Papers. Second International Symposium on Wearable Computers* (pp. 4–8). IEEE.
2. Page, T. (2015). A forecast of the adoption of wearable technology. *International Journal of Technology Diffusion (IJTD)*, *6*(2), 12–29.
3. Roda, A., Guardigli, M., Michelini, E., & Mirasoli, M. (2009). Bioluminescence in analytical chemistry and *in vivo* imaging. *TrAC Trends in Analytical Chemistry*, *28*(3), 307–322.
4. Baldwin, R. R., Cantey, W. E., Maisel, H., & McDermott, J. P. (1956). The optimum strategy in blackjack. *Journal of the American Statistical Association*, *51*(275), 429–439.
5. Siirtola, P., Koskimäki, H., Mönttinen, H., & Röning, J. (2018). Using sleep time data from wearable sensors for early detection of migraine attacks. *Sensors*, *18*(5), 1374.
6. Meisel, C., El Atrache, R., Jackson, M., Schubach, S., Ufongene, C., & Loddenkemper, T. (2020). Machine learning from wristband sensor data for wearable, noninvasive seizure forecasting. *Epilepsia*, *61*(12), 2653–2666.
7. Krijthe, B. P., Kunst, A., Benjamin, E. J., Lip, G. Y., Franco, O. H., Hofman, A., Witteman, J. C., Stricker, B. H., & Heeringa, J. (2013). Projections on the number of individuals with atrial fibrillation in the European Union, from 2000 to 2060. *European Heart Journal*, *34*(35), 2746–2751.
8. Kwon, S., Hong, J., Choi, E. K., Lee, B., Baik, C., Lee, E., Jeong, E. R., Koo, B. K., Oh, S., & Yi, Y. (2020). Detection of atrial fibrillation using a ring-type wearable device (CardioTracker) and deep learning analysis of photoplethysmography signals: A prospective observational proof-of-concept study. *Journal of Medical Internet Research*, *22*(5), e16443.
9. Mei, Z., Gu, X., Chen, H., & Chen, W. (2018). Automatic atrial fibrillation detection based on heart rate variability and spectral features. *IEEE Access*, *6*, 53566–53575.
10. GUSTO Angiographic Investigators. (1993). The effects of tissue plasminogen activator, streptokinase, or both on coronary-artery patency, ventricular function, and survival after acute myocardial infarction. *New England Journal of Medicine*, *329*(22), 1615–1622.
11. Buettner, R., Frick, J., & Rieg, T. (2019). High-performance detection of epilepsy in seizure-free EEG recordings: A novel machine learning approach using very specific epileptic EEG sub-bands. In *ICIS 2019 Dec*.
12. Acharya, U. R., Fujita, H., Oh, S. L., Hagiwara, Y., Tan, J. H., & Adam, M. (2017). Application of deep convolutional neural network for automated detection of myocardial infarction using ECG signals. *Information Sciences*, *415*, 190–198.
13. Hwang, S., Jebelli, H., Choi, B., Choi, M., & Lee, S. (2018). Measuring workers' emotional state during construction tasks using wearable EEG. *Journal of Construction Engineering and Management*, *144*(7), 04018050.
14. Kasabov, N., Scott, N., Tu, E., Marks, S., Sengupta, N., Capecci, E., Othman, M., Doborjeh, M., Murli, N., Hartono, R., & Espinosa-Ramos, J. (2016). Design methodology and selected applications of evolving spatio-temporal data machines in the NeuCube neuromorphic framework. *Neural Networks*, *78*, 1–4.

15. Jaramillo-Yánez, A., Benalcázar, M. E., & Mena-Maldonado, E. (2020). Real-time hand gesture recognition using surface electromyography and machine learning: A systematic literature review. *Sensors*, *20*(9), 2467.
16. Fang, C., Wang, Y., & Gao, S. (2020). Utilizing wearable GRF and EMG sensing system and machine learning algorithms to enable locomotion mode recognition for in-home rehabilitation. In *2020 IEEE International Conference on Flexible and Printable Sensors and Systems (FLEPS)* (pp. 1–4). IEEE.
17. Milosevic, B., Benatti, S., & Farella, E. (2017). Design challenges for wearable EMG applications. In *Design, Automation & Test in Europe Conference & Exhibition (DATE)* (pp. 1432–1437). IEEE.
18. Can, Y. S., Arnrich, B., & Ersoy, C. (2019). Stress detection in daily life scenarios using smartphones and wearable sensors: A survey. *Journal of Biomedical Informatics*, *92*, 103139.
19. Grucza, R., Lecroart, J. L., Carette, G., Hauser, J. J., & Houdas, Y. (1987). Effect of voluntary dehydration on thermoregulatory responses to heat in men and women. *European Journal of Applied Physiology and Occupational Physiology*, *56*(3), 317–322.
20. Un, K. C., Wong, C. K., Lau, Y. M., Lee, J. C., Tam, F. C., Lai, W. H., Lau, Y. M., Chen, H., Wibowo, S., Zhang, X., & Yan, M. (2021). Observational study on wearable biosensors and machine learning-based remote monitoring of COVID-19 patients. *Scientific Reports*, *11*(1), 4388.
21. Castaneda, D., Esparza, A., Ghamari, M., Soltanpur, C., & Nazeran, H. (2018). A review on wearable photoplethysmography sensors and their potential future applications in health care. *International Journal of Biosensors & Bioelectronics*, *4*(4), 195.
22. Akbulut, F. P., Ikitimur, B., & Akan, A. (2020). Wearable sensor-based evaluation of psychosocial stress in patients with metabolic syndrome. *Artificial Intelligence in Medicine*, *104*, 101824.
23. Janidarmian, M., Roshan Fekr, A., Radecka, K., & Zilic, Z. (2017). A comprehensive analysis on wearable acceleration sensors in human activity recognition. *Sensors*, *17*(3), 529.
24. Compagnon, P., Lefebvre, G., Duffner, S., & Garcia, C. (2020). Learning personalized ADL recognition models from few raw data. *Artificial Intelligence in Medicine*, *107*, 101916.
25. Preece, S. J., Goulermas, J. Y., Kenney, L. P., Howard, D., Meijer, K., & Crompton, R. (2009). Activity identification using body-mounted sensors—A review of classification techniques. *Physiological Measurement*, *30*(4), R1.
26. Harris, A., True, H., Hu, Z., Cho, J., Fell, N., & Sartipi, M. (2016). Fall recognition using wearable technologies and machine learning algorithms. In *2016 IEEE International Conference on Big Data (Big Data)* (pp. 3974–3976). IEEE.
27. Liu, K. C., Hsieh, C. Y., Hsu, S. J., & Chan, C. T. (2018). Impact of sampling rate on wearable-based fall detection systems based on machine learning models. *IEEE Sensors Journal*, *18*(23), 9882–9890.
28. Ghasemzadeh, H., Amini, N., Saeedi, R., & Sarrafzadeh, M. (2014). Power-aware computing in wearable sensor networks: An optimal feature selection. *IEEE Transactions on Mobile Computing*, *14*(4), 800–812.

29. Collado-Villaverde, A., Cobos, M., Muñoz, P., & Barrero, D. F. (2020). A simulator to support machine learning-based wearable fall detection systems. *Electronics, 9*(11), 1831.
30. Sundararajan, K., Georgievska, S., Te Lindert, B. H., Gehrman, P. R., Ramautar, J., Mazzotti, D. R., Sabia, S., Weedon, M. N., van Someren, E. J., Ridder, L., & Wang, J. (2021). Sleep classification from wrist-worn accelerometer data using random forests. *Scientific Reports, 11*(1), 24.
31. Cheon, A., Jung, S. Y., Prather, C., Sarmiento, M., Wong, K., & Woodbridge, D. M. (2020). A machine learning approach to detecting low medication state with wearable technologies. In *2020 42nd Annual International Conference of the IEEE Engineering in Medicine & Biology Society (EMBC)* (pp. 4252–4255). IEEE.
32. Amft, O., & Troster, G. (2009). On-body sensing solutions for automatic dietary monitoring. *IEEE Pervasive Computing, 8*(2), 62–70.
33. Bedri, A., Verlekar, A., Thomaz, E., Avva, V., & Starner, T. (2015). A wearable system for detecting eating activities with proximity sensors in the outer ear. In *Proceedings of the 2015 ACM International Symposium on Wearable Computers* (pp. 91–92).
34. Bi, S., Wang, T., Davenport, E., Peterson, R., Halter, R., Sorber, J., & Kotz, D. (2017). Toward a wearable sensor for eating detection. In *Proceedings of the 2017 Workshop on Wearable Systems and Applications* (pp. 17–22).
35. Perez, M. V., Mahaffey, K. W., Hedlin, H., Rumsfeld, J. S., Garcia, A., Ferris, T., Balasubramanian, V., Russo, A. M., Rajmane, A., Cheung, L., & Hung, G. (2019). Large-scale assessment of a smartwatch to identify atrial fibrillation. *New England Journal of Medicine, 381*(20), 1909–1917.
36. Zdravevski, E., Risteska Stojkoska, B., Standl, M., & Schulz, H. (2017). Automatic machine-learning based identification of jogging periods from accelerometer measurements of adolescents under field conditions. *PLOS One, 12*(9), e0184216.
37. Zhang, D., Zhang, H., & Zhao, C. (2022). Prediction of behavior and activity of patients through wearable devices based on deep learning. *Journal of Sensors, 2022*, 1, 2.
38. Gharibans, A. A., Smarr, B. L., Kunkel, D. C., Kriegsfeld, L. J., Mousa, H. M., & Coleman, T. P. (2018). Artifact rejection methodology enables continuous, noninvasive measurement of gastric myoelectric activity in ambulatory subjects. *Scientific Reports, 8*(1), 1, 2.
39. Majaranta, P., & Räihä, K. J. (2002). Twenty years of eye typing: Systems and design issues. In *Proceedings of the 2002 Symposium on Eye Tracking Research & Applications* (pp. 15–22).
40. Tonacci, A., Billeci, L., Di Mambro, I., Marangoni, R., Sanmartin, C., & Venturi, F. (2021). Wearable sensors for assessing the role of olfactory training on the autonomic response to olfactory stimulation. *Sensors, 21*(3), 770.
41. Sabry, F., Eltaras, T., Labda, W., Hamza, F., Alzoubi, K., & Malluhi, Q. (2022). Towards on-device dehydration monitoring using machine learning from wearable device's data. *Sensors, 22*(5), 1887.
42. Abedin, A., Motlagh, F., Shi, Q., Rezatofighi, H., & Ranasinghe, D. (2020). Towards deep clustering of human activities from wearables. In *Proceedings of the 2020 ACM International Symposium on Wearable Computers* (pp. 1–6).

43. Lee, P. T., Chiu, W. C., Ho, Y. H., Tai, Y. C., Lin, C. C., & Lin, C. L. (2021). Development of wearable device and clustering based method for detecting falls in the elderly. In *2021 IEEE 10th Global Conference on Consumer Electronics (GCCE)* (pp. 231–232). IEEE.
44. Reule, S., & Drawz, P. E. (2012). Heart rate and blood pressure: Any possible implications for management of hypertension? *Current Hypertension Reports, 14*, 478–484.
45. Morgenthaler, T., Alessi, C., Friedman, L., Owens, J., Kapur, V., Boehlecke, B., Brown, T., Chesson, A., Coleman, J., Lee-Chiong, T., & Pancer, J. (2007). Practice parameters for the use of actigraphy in the assessment of sleep and sleep disorders: An update for 2007. *Sleep, 30*(4), 519–529.
46. Hirshkowitz, M. (2015). The history of polysomnography: Tool of scientific discovery. In *Sleep Medicine: A Comprehensive Guide to Its Development, Clinical Milestones, and Advances in Treatment* (pp. 91–100).
47. Jiménez, J. F., Illa, F. B., Gil, F. C., Vives, E. C., de Ataur, J. D., Cantoll, J. D., Ortiz, S. L., Trigo, J. M., Canal, J. M., González, M. R., & Santos, J. T. (2007). Resources and delays in the diagnosis of sleep apnea-hypopnea syndrome. *Archivos de Bronconeumología (English Edition), 43*(4), 188–198.
48. Masa, J. F., Corral, J., Pereira, R., Duran-Cantolla, J., Cabello, M., Hernández-Blasco, L., Monasterio, C., Alonso, A., Chiner, E., Zamorano, J., & Aizpuru, F. (2011). Therapeutic decision-making for sleep apnea and hypopnea syndrome using home respiratory polygraphy: A large multicentric study. *American Journal of Respiratory and Critical Care Medicine, 184*(8), 964–971.
49. Sadeh, A. (2011). The role and validity of actigraphy in sleep medicine: An update. *Sleep Medicine Reviews, 15*(4), 259–267.
50. Kim, Y. G., Baltabekova, A. Z., Zhiyenbay, E. E., Aksambayeva, A. S., Shagyrova, Z. S., Khannanov, R., Ramanculov, E. M., & Shustov, A. V. (2017). Recombinant Vaccinia virus-coded interferon inhibitor B18R: Expression, refolding and a use in a mammalian expression system with a RNA-vector. *PLOS One, 12*(12), e0189308.
51. Baldassano, S. N., Brinkmann, B. H., Ung, H., Blevins, T., Conrad, E. C., Leyde, K., Cook, M. J., Khambhati, A. N., Wagenaar, J. B., Worrell, G. A., & Litt, B. (2017). Crowdsourcing seizure detection: Algorithm development and validation on human implanted device recordings. *Brain, 140*(6), 1680–1691.
52. Baumgartner, C., & Koren, J. P. (2018). Seizure detection using scalp-EEG. *Epilepsia, 59*, 14–22.
53. Bagheri, E., Jin, J., Dauwels, J., Cash, S., & Westover, M. B. (2019). A fast machine learning approach to facilitate the detection of interictal epileptiform discharges in the scalp electroencephalogram. *Journal of Neuroscience Methods, 326*, 108362.
54. Clarke, S., Karoly, P. J., Nurse, E., Seneviratne, U., Taylor, J., Knight-Sadler, R., Kerr, R., Moore, B., Hennessy, P., Mendis, D., & Lim, C. (2021). Computer-assisted EEG diagnostic review for idiopathic generalized epilepsy. *Epilepsy & Behavior, 121*, 106556.
55. Seneviratne, U., Boston, R. C., Cook, M., & D'Souza, W. (2017). EEG correlates of seizure freedom in genetic generalized epilepsies. *Neurology: Clinical Practice, 7*(1), 35–44.

56. Beniczky, S., & Ryvlin, P. (2018). Standards for testing and clinical validation of seizure detection devices. *Epilepsia, 59*, 9–13.
57. Poh, M. Z., Loddenkemper, T., Reinsberger, C., Swenson, N. C., Goyal, S., Sabtala, M. C., Madsen, J. R., & Picard, R. W. (2012). Convulsive seizure detection using a wrist-worn electrodermal activity and accelerometry biosensor. *Epilepsia, 53*(5), e93–e97.
58. Onorati, F., Regalia, G., Caborni, C., Migliorini, M., Bender, D., Poh, M. Z., Frazier, C., Kovitch Thropp, E., Mynatt, E. D., Bidwell, J., & Mai, R. (2017). Multicenter clinical assessment of improved wearable multimodal convulsive seizure detectors. *Epilepsia, 58*(11), 1870–1879.
59. Fisher, R. S., Boas, W. V., Blume, W., Elger, C., Genton, P., Lee, P., & Engel, J. (2005). Epileptic seizures and epilepsy: Definitions proposed by the International League Against Epilepsy (ILAE) and the International Bureau for Epilepsy (IBE). *Epilepsia, 46*(4), 470–472.
60. Larsen, S. N., Conradsen, I., Beniczky, S., & Sorensen, H. B. (2014). Detection of tonic epileptic seizures based on surface electromyography. In *2014 36th Annual International Conference of the IEEE Engineering in Medicine and Biology Society* (pp. 942–945). IEEE.
61. Nei, M., Ho, R. T., & Sperling, M. R. (2000). EKG abnormalities during partial seizures in refractory epilepsy. *Epilepsia, 41*(5), 542–548.
62. Meisel, C., & Bailey, K. A. (2019). Identifying signal-dependent information about the preictal state: A comparison across ECoG, EEG, and EKG using deep learning. *EBioMedicine, 45*, 422–431.
63. Ufongene, C., El Atrache, R., Loddenkemper, T., & Meisel, C. (2020). Electrocardiographic changes associated with epilepsy beyond heart rate and their utilization in future seizure detection and forecasting methods. *Clinical Neurophysiology, 131*(4), 866–879.
64. Maimon, O., & Rokach, L. (Eds.). (2005). *Data Mining and Knowledge Discovery Handbook.* Springer.
65. Kohavi, R., & John, G. H. (1997). Wrappers for feature subset selection. *Artificial Intelligence, 97*(1, 2), 273–324.
66. Holden, B., & Gitlesen, J. P. (2006). A total population study of challenging behavior in the county of Hedmark, Norway: Prevalence, and risk markers. *Research in Developmental Disabilities, 27*(4), 456–465.
67. Greco, A., Valenza, G., Lanata, A., Rota, G., & Scilingo, E. P. (2014). Electrodermal activity in bipolar patients during affective elicitation. *IEEE Journal of Biomedical and Health Informatics, 18*(6), 1865–1873.
68. Garcia-Ceja, E., Riegler, M., Nordgreen, T., Jakobsen, P., Oedegaard, K. J., & Tørresen, J. (2018). Mental health monitoring with multimodal sensing and machine learning: A survey. *Pervasive and Mobile Computing, 51*, 1–26.
69. An, Q., Gan, S., Xu, J., Bao, Y., Wu, T., Kong, H., Zhong, L., Ma, Y., Song, Z., & Niu, L. (2019). A multichannel electrochemical all-solid-state wearable potentiometric sensor for real-time sweat ion monitoring. *Electrochemistry Communications, 107*, 106553.

70. Alvarez, A., Severeyn, E., Velásquez, J., Wong, S., Perpiñan, G., & Huerta, M. (2019). Machine learning methods in the classification of athletes' dehydration. In *2019 IEEE Fourth Ecuador Technical Chapters Meeting (ETCM)* (pp. 1–5). IEEE.
71. Aggarwal, R., & Das, M. L. (2012). RFID security in the context of "internet of things." In *Proceedings of the First International Conference on Security of Internet of Things* (pp. 51–56).
72. Heidt, S. T., Kratz, A., Najarian, K., Hassett, A. L., Oral, H., Gonzalez, R., Nallamothu, B. K., Clauw, D., & Ghanbari, H. (2016). Symptoms in atrial fibrillation: A contemporary review and future directions. *Journal of Atrial Fibrillation, 9*(1).
73. Tuli, S., Basumatary, N., Gill, S. S., Kahani, M., Arya, R. C., Wander, G. S., & Buyya, R. (2020). HealthFog: An ensemble deep learning-based smart healthcare system for automatic diagnosis of heart diseases in integrated IoT and fog computing environments. *Future Generation Computer Systems, 104*, 187–200.
74. Glasziou, P., Altman, D. G., Bossuyt, P., Boutron, I., Clarke, M., Julious, S., Michie, S., Moher, D., & Wager, E. (2014). Reducing waste from incomplete or unusable reports of biomedical research. *The Lancet, 383*(9913), 267–276.
75. Keerthy, A. S., & Priya, S. M. (2021). Artificial Intelligence in healthcare databases. In *Artificial Intelligence in Healthcare Databases* (pp. 19–34). Springer International Publishing.
76. Nelson, B., Low, C., Jacobson, N., Arean, P., Torous, J., & Allen, N. (2020). Guidelines for wrist-worn wearable assessment of heart rate in biobehavioral research. *Nature, 3*.
77. Kim, J. K., Jung, H., Hwang, G. B., Gwon, O. S., & Lee, S. E. (2020). Design of low-power SoC for wearable healthcare device. *Journal of Circuits, Systems, and Computers, 29*(6), 2050085.
78. Nizam, Y., & Jamil, M. M. (2020). Classification of daily life activities for human fall detection: A systematic review of the techniques and approaches. In *Challenges and Trends in Multimodal Fall Detection for Healthcare* (pp. 137–179).
79. Faustino, M., Calado, J., Sarraipa, J., & Jardim-Gonçalves, R. (2022). Adaptable power consumption profiles for wearable localization devices. In *Proceedings of the SEIA'2019 Conference Proceedings* (p. 200). Tenerife, Canary.
80. Rahmani, A. M., Gia, T. N., Negash, B., Anzanpour, A., Azimi, I., Jiang, M., & Liljeberg, P. (2018). Exploiting smart e-health gateways at the edge of healthcare Internet-of-Things: A fog computing approach. *Future Generation Computer Systems, 78*, 641–658.
81. Lewandowski, M., Płaczek, B. O., & Bernas, M. (2021). Classifier-based data transmission reduction in wearable sensor network for human activity monitoring. *Sensors, 21*(1).
82. Zhou, J., & Wang, C. (2017). An ultra-low power turning angle based biomedical signal compression engine with adaptive threshold tuning. *Sensors, 17*, 1809.
83. Sherratt, R. S., Janko, B., Hui, T., et al. (2019). Task scheduling to constrain peak current consumption in wearable healthcare sensors. *Electronics, 8*(7).
84. Varshney, U. (2007). Pervasive healthcare and wireless health monitoring. *Mobile Networks and Applications, 12*, 113–127.
85. Scrugli, M. A., Loi, D., Raffo, L., & Meloni, P. (2019). A runtime-adaptive cognitive IoT node for healthcare monitoring. In *Proceedings of the 16th ACM International*

Conference on Computing Frontiers, CF '19 (pp. 350–357). Association for Computing Machinery.
86. Takao, T. (2016). Design consideration of sustainable self-sufficient energy system using ferroelectric material for wearable healthcare. *International Journal of Energy Research, 40*(15), 2176–2186.
87. Zhang, M., & Sawchuk, A. (2012). A feature selection-based framework for human activity recognition using wearable multimodal sensors. In *6th International ICST Conference on Body Area Networks* (pp. 1–5).
88. Mairittha, N., Mairittha, T., & Inoue, S. (2020). On-device deep personalization for robust activity data collection. *Sensors, 21.*
89. Lara, O. D., & Labrador, M. A. (2012). A survey on human activity recognition using wearable sensors. *IEEE Communications Surveys & Tutorials, 15*(3), 1192–1209.
90. Ferrari, A., Micucci, D., Mobilio, M., & Napoletano, P. (2022). Deep learning and model personalization in sensor-based human activity recognition. *Journal of Reliable Intelligent Environments, 54.*
91. Zhang, T., Lu, J., Hu, F., & Qi, H. (2014). Bluetooth low energy for wearable sensor-based healthcare systems. In *Proceedings of the 2014 IEEE Healthcare Innovation Conference (HIC)* (pp. 251–254). Seattle, USA.
92. Prance, H. (2011). Sensor developments for electrophysiological monitoring in healthcare. *Applied Biomedical Engineering*, 265–286.
93. Adi, E., Anwar, A., & Baig, Z. (2020). Machine learning and data analytics for the IoT. *Neural Computing & Applications, 32*, 16205.
94. Alsubaei, F., Abuhussein, A., & Shiva, S. (2017). Security and privacy in the internet of medical things: Taxonomy and risk assessment. In *Proceedings of the 2017 IEEE 42nd Conference on Local Computer Networks Workshops (LCN Workshops)* (pp. 112–120). Singapore.

CHAPTER 6

Prediction of Diabetes Using Machine Learning

ABSTRACT

Diabetes threatens worldwide health, making it a major global health issue. This metabolic illness causes hyperglycemia, cardiovascular disease, renal failure, and neuropathy. Scientists have been working on a reliable diabetes prediction algorithm for a long time. Major research obstacles can only be overcome using big data analytics and machine learning-based methodologies due to a dearth of adequate data sets and modeling techniques. Machine learning is an emerging data science area that investigates automatic learning through practice and observation. This study combines machine learning results to enhance diabetes prediction. Using machine learning, this study aims to predict diabetes. Researchers hope to provide a method for early, reliable diabetes detection.

6.1 INTRODUCTION

The occurrence of chronic diseases on a large scale has become a worldwide problem, impacting both developing and developed countries. Diabetes mellitus is a prevalent and incapacitating condition that significantly reduces people's lifespan worldwide. Countries located within the visible light spectrum, such as India and certain regions of the Sahara, are experiencing a significant rise in the prevalence of diabetes. The consequences of diabetes are expected to be severe and persistent. Numerous global organizations and firms are allocating funds to conduct research aimed at enhancing the early detection and treatment of diabetes, with the ultimate goal of mitigating the worldwide mortality rate associated with this

Machine Learning in Healthcare: Advances and Future Prospects. Rishabha Malviya, Niranjan Kaushik, Tamanna Rai, M. P. Saraswathy, and Rajendra Awasthi (Authors)

condition. Insufficient nourishment and a sedentary lifestyle both elevate the risk of diabetes; however, individuals who allocate time for consistent physical activity frequently see favorable alterations in their condition. The human body is sensitive to several harmful effects of diabetes, which can affect organs such as the brain, eyes, kidneys, and nervous system [1].

In recent years, diabetes has emerged as a significant contributor to mortality in countries with low per capita income. Both the government and private sectors are providing financial support for the discovery of a drug or cure for this lethal disease. Insulin resistance, the underlying factor of diabetes, leads to persistently high levels of blood sugar in individuals. Diabetic patients have an impaired ability to convert dietary carbohydrates into glucose, which is essential for their daily energy needs. Consequently, the level of glucose in the bloodstream progressively increases. This signifies impaired distribution of glucose to all cells in the body, resulting in its accumulation in the bloodstream [2]. Low-carbohydrate diets have been suggested to be beneficial in the treatment of diabetes. Utilizing a diverse range of predictive, quantitative, and statistical models is essential for diagnosing diseases. Diabetes is linked to a higher risk of stroke and cardiovascular disease, in addition to accelerating the onset of other medical conditions. Diabetic individuals experience impaired cellular function, which results in weight gain [3]. The goal is to enhance the application of big data analytics and machine learning algorithms by healthcare providers in decision-making, disease prediction, and prognosis [4].

6.2 DIABETIC MELLITUS

Diabetes is one of the leading causes of death globally. There are a lot of complications that can arise from this disease, including heart issues, loss of vision, renal failure, and many more. In the past, patients had to make an appointment with a doctor at a diagnostic center, wait at least a day, and then get their results. In addition, they are required to continue making payments to access their diagnosis report, even though it provides no benefit to them. Insulin insufficiency leads to elevated blood glucose levels and impaired metabolism of carbohydrates, lipids, and proteins. Diabetes, encompassing both type 1 and type 2, belongs to a category of metabolic disorders characterized by significant disruptions in insulin secretion and/or function. Diet, sedentary behavior, and genetic predisposition are associated with type 1 diabetes, whereas the immune system's destruction of pancreatic beta cells in the Langerhans islets is a hallmark of type 2 diabetes. Type 2 diabetes is a prevalent endocrine disorder that impacts 200 million people globally. Projections on the future prevalence of diabetes are alarming. Type 2 diabetes, characterized by insulin resistance, accounts for 90% of all diabetes cases. Diabetes mellitus can be classified into various types based on differences in insulin secretion profiles and/or ages of onset. These types include MODY (maturity-onset diabetes of the young), mitochondrial diabetes, neonatal diabetes, and gestational diabetes. MODY is a type of monogenic diabetes first described as a mild and asymptomatic form of diabetes that was observed in non-obese children, adolescents, and young adults. The symptoms of diabetes mellitus include polyuria, polydipsia, and anorexia. Plasma glucose levels exceeding 7.0 mmol/L indicate the presence of diabetes.

6.3 MACHINE LEARNING

Describing machine learning is challenging because of its extensive and diverse nature, encompassing various disciplines including statistics, algebra, data processing, and knowledge analytics [5]. Machine learning is a subfield of artificial intelligence (AI) that uses past data instances to enhance its performance in upcoming instances. The goal of machine learning is to facilitate the development of adaptable and continuously upgraded software. A computer program can acquire knowledge from experience E if and only if it enhances its performance on tasks in class T, as evaluated by performance measure P, following exposure E. The development of machine learning techniques has allowed us to create a system that uses data mining to determine whether a certain patient has diabetes or not. The ability to foresee the progression of a disease allows patients to receive treatment far before their conditions worsen to a critical stage. Using data mining, insights can be uncovered in a sea of information on diabetes. Because of this, diabetes research is more crucial than ever [6]. Learning can be achieved through three distinct methods: supervised, unsupervised, and semi-supervised.

6.3.1 INSTRUCTED HUMAN PREDICTIVE MODELS

These forecasting models are built using supervised learning techniques. A predictive model can identify the missing value by analyzing the existing data and making informed assumptions. For a supervised learning system to accurately forecast the performance of a new dataset, it necessitates the presence of input and output data examples. Supervised learning encompasses several techniques, such as decision trees (DTs), Bayesian approaches, artificial neural networks (ANNs), instance-based learning (IBL), and ensemble methods. The efficacy of machine learning can be directly linked to these methodologies [7].

6.3.2 LEARNING WITHOUT SUPERVISION OR DESCRIPTIVE MODELS

Unsupervised learning is used to generate descriptive models. The result of this model is uncertain, even though the inputs are known. Unsupervised learning is frequently employed for analyzing transactional data. The strategy utilizes clustering algorithms such as k-means and k-medians.

6.3.3 LEARNING WITH LIMITED SUPERVISION

In semi-supervised learning, the training dataset consists of both labeled and unlabeled examples. Semi-supervised learning encompasses classification and regression methods that rely on minimal human input. Regression techniques include logistic regression (LR) and linear regression.

To identify people who are at the greatest risk of acquiring diabetes, there is an urgent need for additional investigation and the development of more accurate methods. We must develop a system that is based on three distinct classification methods: naive bayes (NB), support vector machine (SVM), and decision stump. These approaches have the potential to predict the outcomes of LR and ANN algorithms. Mining, a topic that evolved a great deal later than machine learning, is significantly influenced by the constitutional standards that corporations must adhere to ensure compliance with data science. In statistical terms, cluster analysis, which is also commonly referred to as clustering, is a method for organizing data that involves locating groupings of objects within the data that are more similar to one another than they are to those in other clusters. Exploratory data mining has this as one of its primary goals, and it also finds applications in a wide variety of other fields, including machine learning, image processing, pattern recognition, bioinformatics, information retrieval, data compression, and the design of graphical user interfaces.

6.4 DESIGNING A 5G METHOD FOR CONTROLLING DIABETES PATIENTS

The monitoring of diabetic patients involves the utilization of sensors, wearables, a smartphone application, a server that has a database, and 5G networks. There are sensors on the mobile device that are connected wirelessly. 5G technology, on the other hand, can connect mobile devices to the cellular network and send data to the primary data storage center. The blood glucose levels, body temperature, and activity levels of diabetes patients would be monitored by a smartphone that is connected to a 5G network. The data would then be transmitted to a designated base station. Subsequently, technology utilizes AI and machine learning to assist individuals in maintaining their glucose levels and predicting changes in their health. Since 5G technology can support over 60,000 connections with low latency, we can utilize it for remote patient monitoring. The sensors, the data gathering, the transmission, and the database were the four fundamental tiers that we presented. The representation of the patient diabetes monitoring system is shown in Figure 6.1.

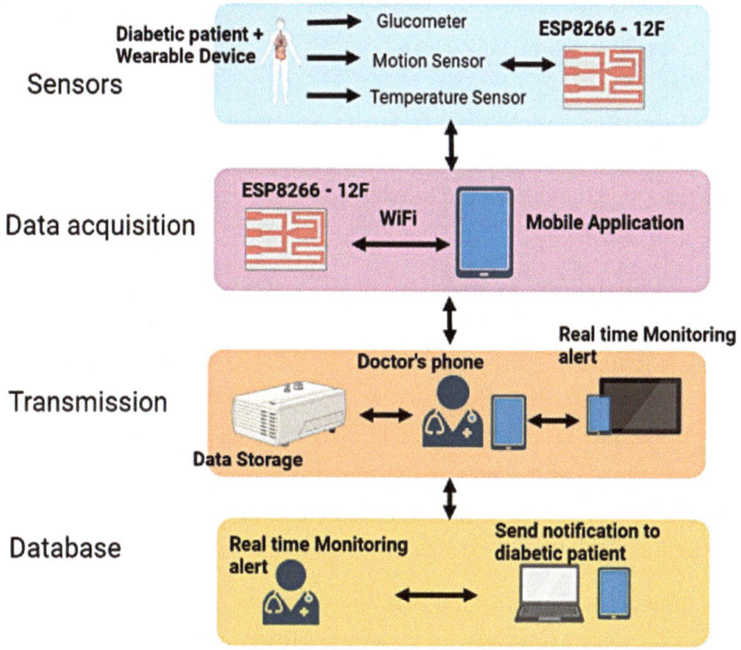

FIGURE 6.1 Representation of various components of a patient diabetes monitoring system.

6.4.1 SENSORS

The sensors for measuring blood glucose, temperature, and movement are all positioned in this location. This layer incorporates the ESP8266 module, which establishes connections with the sensors and enables a wireless interface for relaying data to the patient's smartphone. Hence, the sensors are accountable for gathering and transmitting the data to the patient's smartphone.

6.4.2 DATA ACQUISITION LAYER

The patient's smartphone and the application that collects data are stored at this location. The mobile application provides sensor data. Additionally, the 5G network is capable of transmitting data to the base station, which enables several connections to be made simultaneously within the coverage area. The goal is to achieve 1 million devices per kilometer, which is 10 times more than the current 4G standard.

6.4.3 TRANSMISSION LAYER

The information is delivered from the smartphone to the database, where it is processed, and then it is sent to the mobile device of the doctor for examination. This procedure takes place over 5G.

6.4.4 DATABASE LAYER

The sensor readings are recorded in this data processing module before being sent to one of the many AI algorithms that are currently accessible for analysis and categorization. Utilizing machine learning techniques, the server ascertains if the data is positive (also known as true positive (TP)) or negative (also known as false negative (FN)). If something atypical occurs, the system will automatically provide a warning. The doctors receive notifications on their mobile devices, review them, and then convey their advice and treatments to their patients, which the patients can access on their own mobile devices [8–10].

6.5 PROPOSED SYSTEM

Classification is a powerful tool for addressing numerous practical challenges in the real world. Increasing the number of samples does not necessarily lead to improvement in many categorization situations. Algorithms exhibit rapid but inadequate data classification. Enhancing classification accuracy can be achieved by conducting tests on a reduced sample size while training the model on a larger dataset. The most effective methods for predicting diabetes are the naive net, ANN, SVM, decision stump, K-nearest neighbor (KNN), and LR.

6.5.1 *K-NEAREST NEIGHBOR (KNN)*

K-nearest neighbor (KNN) is a widely used technique in machine learning as a form of supervised learning. Primarily, it aids in resolving issues related to categorization. KNN uses the average distance between each object and every other object in the training data to classify the data. KNNs are employed to ascertain the classification of an object. Before executing the method, a positive integer *K* is established. The Euclidean distance is commonly employed for converting measurements between length, area, volume, and area-based measuring systems [12]. The Euclidean approach provides the following formula for calculating distance:

$$n = \sqrt{\sum_{i=1}^{k} (x_i - y_i)^2}$$

$$ttan = \sum_{i=1}^{k} (x_i - y_i)$$

6.5.2 *ARTIFICIAL NEURAL NETWORK (ANN)*

The artificial neural network (ANN), like a human brain, can execute a diverse range of intricate activities. Multi-layered or cube-shaped ANNs have demonstrated effectiveness since signals propagate from the front to the rear. Sometimes, backpropagation is utilized during training when the front neuronal units are reset through forward stimulation once the intended outcome is already established. Contemporary networks have increased disorganization and intricate connections, with activation and inhibition displaying heightened dynamism. Neural networks that lack flexibility restrict their potential by impeding the development of new connections or brain units, as well as the inhibition of existing ones based on regulations. The input layer, hidden layer, and output layer all have significant functions in a well-constructed ANN. The input neurons define all the attribute values utilized by the data mining model. Seven neurons corresponding to the seven variables (skin thickness, blood sugar, blood pressure, body mass index, diabetes pedigree function, insulin, and age) linked with each data point can be used.

Neurons in the hidden layer receive input from the input neurons and transmit it to the output neurons in the same layer. Probabilities are evaluated in the hidden layer. Each input to the hidden neuron is assigned a different weight to illustrate how significant it is. The ability of an activation function to produce a progressive change in output in response to changes in the value of the input is among the most essential characteristics of this type of function. Pattern recognition has found applications in a wide variety of domains, including sequence identification, finance, data mining, medical diagnosis, visualization, and the classification of spam in electronic mail. Other areas include system identification and control, quantum chemistry, gaming, and decision-making. Rather than collecting data from a single patient, data from a large group of patients can be used to construct more specific models. It is not necessary to assume any relationships between the variables to apply any of the models. Neural network predictions for colorectal cancer have also been documented. Neural networks have the potential to provide more accurate predictions for the likelihood of recovery for patients with colorectal cancer compared to the existing standard of care. Following the training process, the networks demonstrated the capability to provide precise prognostications regarding the well-being of several patients across multiple establishments [13].

6.5.3 SUPPORT VECTOR MACHINE (SVM)

The support vector machine (SVM) algorithm is the most often employed machine learning method. The first phase in the SVM process involves determining the appropriate hyperplane, which is then followed by the process of increasing the distances between data points that are near one another. The problem is solved by adding a feature with the equation $z =$

$x^2 + y^2$ to the SVM. The data is categorized into meaningful groups using a SVM classifier [14].

6.5.4 DECISION STUMP

It is one of the most popular classification algorithms for determining unique attribute values in the field of machine learning. A decision stump is a simple machine learning model of a one-level DT. Decision stumps form the foundation of complicated ensemble learning algorithms like AdaBoost for diabetic disease prediction. Decision stumps are simple models that make one-feature decisions. Decision stumps can be merged into an ensemble like AdaBoost to generate a more powerful and accurate predictive model. AdaBoost sequentially trains several decision stumps, giving more weight to cases misclassified by earlier stumps. This improves model prediction. The final prediction is a weighted mixture of decision stumps. The technique is effective for imbalanced datasets or complex feature-target variable connections since AdaBoost can adapt to these patterns during training [15].

6.5.5 NAIVE NETS

In addition to having a relatively low level of temporal complexity, this method makes use of the probability formula to carry out computations that are dependent on the possibility of incidents occurring. In the past, it was utilized to maximize the probability of a class or feature, where (C|F) is equal to the PR (class | feature). After the data has been transformed into a frequency table, the next step is to determine the likelihood of the information being presented. In the end, during the process of prediction, the Naïve Bayes equation is utilized.

6.5.6 LOGISTIC REGRESSION (LR) FRAMEWORK

As a result of the parameterizable nature of machine learning models, hyperparameters can be modified, and the behavior of these models can be fine-tuned in response to alterations in the issue statement. When describing a model, one can utilize a wide variety of traits and qualities, respectively. Finding an appropriate combination of attributes can be viewed as a search problem, which is a perspective that is conceivable to take. The tuning of the LR has been accomplished using both a grid search and a random search. One method that may be utilized to ascertain the sample distribution of all parameters and the required number of iterations to locate the optimal model is the random search. It is possible to calculate the value of a hyperparameter by taking the average of many samples. A grid search technique can achieve this.

$$\underset{wb}{\text{Min}} = \left(-\frac{1}{2} \sum_{i=1}^{n} \Pr(y_i = 1 \mid x_i : w,\, b) + R(w) \right)$$

$$R(w) = \left(a \sum_{i=1}^{p} \mid w \mid + \frac{(1-\alpha)}{2} \mid w \mid T \mid L \mid w \mid \right)$$

The grid search procedure breaks down each individual parameter and looks for its sweet spot while the others are held steady. This is because the model score experienced a decrease in its predictive ability. It possesses a significantly higher capacity for exploration compared to a random search. The robustness in the crucial area enables it to select the optimal configuration (hyperparameter). Increasing the predictive efficacy and efficiency of the LR classifier can be accomplished through the utilization of grid search [16].

6.6 CONCLUSION

Machine learning has the potential to completely revolutionize the field of diabetes risk prediction. This is due to the availability of cutting-edge computational methods, an abundance of epidemiological and genetic diabetes risk datasets, and the capability of machine learning to fundamentally transform the field. All these factors have contributed to the potential for machine learning to completely change the field. Timely diagnosis is crucial for optimal diabetes care. This chapter outlines machine learning techniques for predicting blood sugar levels in diabetic patients. Researchers can utilize this approach to create a reliable tool that will help doctors make better-informed decisions about the patient's condition.

KEYWORDS

- **artificial neural networks**
- **blood sugar levels**
- **cardiovascular disease**
- **diabetes mellitus**
- **logistic regression**
- **machine learning**
- **supervised learning**

REFERENCES

1. Gomes, J. M., de Assis Costa, J., & Alfenas, R. D. (2017). Metabolic endotoxemia and diabetes mellitus: A systematic review. *Metabolism, 68*, 133–144.
2. Bishop, C. M., & Nasrabadi, N. M. (2006). *Pattern Recognition and Machine Learning*. Springer.
3. Massari, H. E., Mhammedi, S., Sabouri, Z., & Gherabi, N. (2022). Ontology-based machine learning to predict diabetes patients. In *Advances in Information, Communication and Cybersecurity: Proceedings of ICI2C'21* (pp. 437–445).
4. Mahabub, A. (2019). A robust voting approach for diabetes prediction using traditional machine learning techniques. *SN Applied Sciences, 1*(12), 1667.
5. Larabi-Marie-Sainte, S., Aburahmah, L., Almohaini, R., & Saba, T. (2019). Current techniques for diabetes prediction: Review and case study. *Applied Sciences, 9*(21), 4604.
6. Devi, M. R., & Shyla, J. M. (2016). Analysis of various data mining techniques to predict diabetes mellitus. *International Journal of Applied Engineering Research, 11*(1), 727–730.
7. Nithya, B., & Ilango, V. (2017). Predictive analytics in healthcare using machine learning tools and techniques. In *2017 International Conference on Intelligent Computing and Control Systems (ICICCS)* (pp. 492–499). IEEE.
8. Tepla, T. L., Izonin, I. V., Duriagina, Z. A., Tkachenko, R. O., Trostianchyn, A. M., Lemishka, I. A., Kulyk, V. V., & Kovbasyuk, T. M. (2018). Alloys selection based on the supervised learning technique for design of biocompatible medical materials. *Archives of Materials Science and Engineering, 1*(93), 32–40.
9. Tkachenko, R., Doroshenko, A., Izonin, I., Tsymbal, Y., & Havrysh, B. (2019). Imbalance data classification via neural-like structures of geometric transformations model: Local and global approaches. In *Advances in Computer Science for Engineering and Education 13* (pp. 112–122). Springer International Publishing.

10. Reinhardt, A. (2004). A Machine-To-Machine 'Internet of Things.' *Business Week, 3880*, 102.
11. Khambra, G., & Shukla, P. (2021). Novel machine learning applications on fly ash-based concrete: An overview. *Materials Today: Proceedings, 80*(3), 3411–3417.
12. Yu, Y., Lin, H., Meng, J., Wei, X., Guo, H., & Zhao, Z. (2017). Deep transfer learning for modality classification of medical images. *Information*, *8*(3), 91.
13. Giri, D., Acharya, U. R., Martis, R. J., Sree, S. V., Lim, T. C., VI, T. A., & Suri, J. S. (2013). Automated diagnosis of coronary artery disease affected patients using LDA, PCA, ICA and discrete wavelet transform. *Knowledge-Based Systems*, *37*, 274–282.
14. Karatsiolis, S., & Schizas, C. N. (2012). Region-based Support Vector Machine algorithm for medical diagnosis on Pima Indian Diabetes dataset. In *2012 IEEE 12th International Conference on Bioinformatics & Bioengineering (BIBE)* (pp. 139–144).
15. Patil, B. M., Joshi, R. C., & Toshniwal, D. (2010). Hybrid prediction model for type-2 diabetic patients. *Expert Systems with Applications*, *37*(12), 8102–8108.
16. Roy, V., Shukla, S., Shukla, P. K., & Rawat, P. (2017). Gaussian elimination-based novel canonical correlation analysis method for EEG motion artifact removal. *Journal of Healthcare Engineering, 2017.*

CHAPTER 7

Mental Health Index Management Using Machine Learning

ABSTRACT

The increasing frequency of mental health concerns and the need for more efficient healthcare have inspired an investigation into the integration of machine learning techniques in the treatment of mental health disorders. This chapter presents a contemporary systematic evaluation of various machine learning algorithms employed in the prediction of mental health issues. Additionally, this discussion will encompass the obstacles, constraints, and potential advancements associated with the utilization of machine learning in the context of mental health.

7.1 INTRODUCTION

The rapidly increasing prevalence of psychological disorders is largely due to the dynamic social dynamics of today. The World Health Organization (WHO) defines mental health as "the degree to which an individual can deal successfully with the ordinary stressors of life, to build and sustain rewarding relationships both at work and in one's personal life, and to give back to one's community" [1]. A person's lifestyle and surroundings can have a detrimental impact on their mental health in several ways, including financial difficulties, work-related stress, marital problems, family issues, and violent acts [2]. The aforementioned factors can further exacerbate psychological disorders like depression, anxiety, and stress, which have a significant negative impact on a person's overall health and quality of

Machine Learning in Healthcare: Advances and Future Prospects. Rishabha Malviya, Niranjan Kaushik, Tamanna Rai, M. P. Saraswathy, and Rajendra Awasthi (Authors)

life. Mental illness accounts for approximately 13% of the global burden of diseases, affecting around 450 million people [3]. Based on data from the WHO, it is estimated that approximately 25% of individuals may experience a mental health condition at some stage throughout their lifespan [4]. The WHO announced recommendations in 2018 aimed at enhancing the overall physical fitness of individuals with severe mental health disorders [5]. A global population of around 350 million individuals experiences depression, a condition that can lead to the manifestation of suicidal thoughts and actions [6]. As a result, the WHO anticipated in its Comprehensive Mental Health Action Plan that people suffering from mental illnesses would be able to rehabilitate and lead normal lives [7]. Early detection and implementation of a treatment regimen are crucial for addressing mental health concerns. Early identification, precise diagnosis, and effective intervention can be advantageous for individuals with mental health challenges [8]. The effects of mental illness can be catastrophic for the afflicted individual, their loved ones, and the community at large. Conventional methodologies to detect mental health conditions generally involve the utilization of questionnaires, self-report measures, or face-to-face interviews. These are time-consuming and labor-intensive methods [9]. Previous studies have employed smartphones and wearable sensors to facilitate healthcare and mental health monitoring. However, these technologies have predominantly been utilized for individuals who have already received a mental disorder diagnosis and are consequently subject to intensive observation [10–13].

7.2 MACHINE LEARNING IN MENTAL HEALTH INDEX

Machine learning, a subfield of artificial intelligence (AI), encompasses three primary domains of investigation: categorization, regression, and clustering. The use of data and algorithms facilitates the ability to gain knowledge and improve performance in a way that is similar to human behavior [14]. Machine learning has demonstrated effectiveness in various domains within the field of psychology and holds significant potential for the diagnosis and treatment of mental health disorders, as well as other health-related consequences. These algorithms require a substantial amount of data to acquire knowledge of patterns and perform classification. The utilization of supervised learning for the prediction of psychological disorders has gained significant popularity. Supervised learning involves the process of acquiring knowledge by establishing a connection between input parameters and a target variable, with the ultimate goal of accurately predicting new data [15]. The support vector machine (SVM) is an example of a supervised learning algorithm, as it can be utilized for both classification and regression tasks. The process of margin computation involves the partitioning of an n-dimensional space into distinct classes based on the placement of a hyperplane, which represents the optimal decision boundary. The process of cataloging newly acquired information is crucial. SVMs demonstrate exceptional performance in the processing of both semi-structured and fully structured data. The duration of model training increases proportionally with the size of the dataset. The effectiveness of it is diminished. SVMs have limitations when used with datasets that contain noise. Decision trees (DTs) can perform various tasks, such as classification, regression, and supervised learning. By employing a piecewise constant approximation, it is possible to accurately represent a tree. The utilization of input data attributes enables the acquisition of decision rules, which are subsequently employed to forecast the values of target variables. The range of logistic regression (LR) predictions is not restricted to the values of 0 and 1 due to the categorical nature of the dependent variable. The naive bayes (NB) algorithm offers a probabilistic approach to classification by utilizing Bayes' theorem. The proposed model advocates for the examination of individual characteristics within each dataset. Deep learning (DL) is a specialized field within the broader domain of machine learning [16]. DL is a specialized area within machine learning that focuses on the automatic identification of distinct characteristics associated with different types of input, such as raw textual information and images, without the need for preprocessing. Due to the autonomous learning capabilities of the architecture, a substantial volume of data is employed to provide feature acquisition and performance enhancements without the need for human intervention. Consequently, there has been a recent increase in scholarly endeavors dedicated to enhancing the utilization of DL techniques to detect and diagnose mental disorders. The utilization of AI in the field of medical care has prompted several studies and articles that investigate the potential applications of machine learning and DL in enhancing our comprehension of various health concerns. The utilization of AI in the field of medical research has witnessed significant growth, mostly driven by the urgency and significance of the task at hand.

This expansion has encompassed the realm of mental health, where AI is now being employed to identify various mental health disorders [17]. Recent advancements in machine learning have demonstrated considerable promise in the domain of mental health diagnosis. An example of a notable advancement is the ability to seamlessly interface with electronic health records (EHRs). In recent times, there has been a noticeable increase in the utilization of data analysis from EHRs to aid in the identification and assessment of mental health disorders. Wearable devices, such as smartwatches and activity trackers, offer a substantial amount of data that can be subjected to analysis through machine learning algorithms. This approach has the potential to facilitate continuous monitoring of mental well-being and the early detection of disorders. Predictive modeling is a discipline that incorporates machine learning techniques. Machine learning algorithms possess the capability to identify individuals who may be susceptible to developing mental health disorders. Consequently, the timely treatment of mental health issues can effectively prevent the development of substantial psychological disorders. Machine learning has been employed to develop automated screening methods for a diverse range of mental health disorders.

Machine learning is employed within this field of psychiatry to facilitate the detection and recognition of mental disorders by analyzing certain patterns present in patient data. Various sources can provide data for research purposes in the medical industry, including records, brain magnetic resonance imaging (MRI) images, and social media posts. To perform this task, a variety of algorithms are employed, including both supervised algorithms that require labeled data for training and unsupervised algorithms that can autonomously identify patterns within the data. With enough information, a model can be trained to predict whether a certain individual is experiencing a particular mental health illness and, if so, to what extent. The forecast is generated by employing machine learning techniques, wherein novel data is input into computational models and the outcomes are utilized to inform their evaluation [18].

7.3 TYPES OF MENTAL HEALTH PROBLEMS

Individuals afflicted with mental illness may encounter impairments in their cognitive processes, emotional experiences, and behavioral responses. The presence of mental health concerns has the potential to hinder a child's ability to acquire knowledge and skills. Moreover, individuals who have mental health conditions impose a significant strain on their family members, colleagues, and broader society. Mental disorders include schizophrenia, depression, bipolar disorder, and anxiety.

7.3.1 DEPRESSION

The primary indicator of depression is characterized by profound distress, as it is the emotional state that experiences the most pronounced impact. In certain instances, it is possible that symptoms associated with depression, such as anger, impatience, and loss of interest, may assume a position of greater significance. Physiological manifestations encompass difficulties with sleep, an inability to retain food, and overall fatigue. Cognitive manifestations involve impaired cognitive functioning, contemplation of suicide, and experiences of guilt. The recurrence of depressive episodes is a common phenomenon observed in individuals diagnosed with depression [18]. Many depressed individuals may never fully recover from their illness, instead establishing a chronic, moderate form of it [19].

7.3.2 SCHIZOPHRENIA

Schizophrenia is characterized by the occurrence of psychotic episodes, which manifest as hallucinations and delusions. The occurrence of hallucinations is inherently subjective, making it challenging to provide a comprehensive explanation due to their idiosyncratic nature. In contrast, individuals diagnosed with delusions exhibit beliefs that are incongruent with the actual world. Symptoms commonly associated with schizophrenia include social withdrawal, heightened rage, and a general escalation in atypical behavior. Investigations are currently being conducted to determine whether the early recognition of such indicators and care could improve results [20].

7.3.3 ANXIETY

Another prevalent mental health issue is anxiety, which is characterized by excessive worry about unimportant things. The manifestation of panic disorders is marked by physiological symptoms such as increased heart rate, perspiration, and dizziness, which are believed to originate from sudden and unpredictable panic attacks and intense terror. Generalized anxiety disorder is characterized by a tendency toward excessive worrying. Post-traumatic stress disorder is known to induce a state of emotional numbing after experiencing a traumatic event. Numerous individuals afflicted with social anxiety experience profound distress when confronted with large gatherings. According to survey data, a significant number of individuals with anxiety-related issues tend to postpone seeking appropriate medical intervention for an extended period [21].

7.3.4 MANIC AND DEPRESSION

Manic and depressive episodes are considered essential diagnostic criteria for bipolar disorder, which is classified as a distinct form of mental illness. It is conceivable to experience a manic or depressive episode. Mania is distinguished by heightened degrees of restlessness, physical movement, and drive, accompanied by a reduced requirement for sleep. Individuals experiencing mania often exhibit behaviors that are characterized by a propensity for engaging in risky activities. On the other hand, the manifestations of a depressive episode in individuals with bipolar disorder show a striking resemblance to those observed in cases of depression. Restoration of pre-episode functioning has been reported, while a significant number of patients continue to experience persistent debilitating symptoms even after the conclusion of an episode [22].

7.4 MACHINE LEARNING AND DEEP LEARNING (DL) METHODOLOGIES APPLIED

There are numerous methodologies and protocols for predicting mental diseases. These methodologies find application in the field of AI, encompassing machine learning, deep neural networks (DNNs), and even robotics. The main objectives of these methodologies are the determination of the underlying etiology of these disorders, their precise diagnosis, and the prediction of their therapeutic outcomes. The 2019 Open-Source Mental Illness Survey suggested that identifying and leveraging factors that negatively impact the mental well-being of employees in both technical and non-technical contexts could serve as predictors [23]. In a further study conducted by Katarya et al., machine learning techniques were employed to uncover factors related to the COVID-19 pandemic that can predict emotional distress [24]. Given the prevailing global health crisis, the sections present an analysis highlighting the significance of assessing physiological manifestations of emotional distress and coping mechanisms. The aim is to enhance the knowledge base for mental health examinations and treatments.

7.4.1 DETECTION OF BIPOLAR DISORDER

7.4.1.1 EEG FEATURES

EEG biomarkers are used in studies to help diagnose bipolar disorder. Before developing the training model, Erguzel et al. conducted two trials using different feature selection methodologies [25]. As a quantitative biomarker of EEG activity, EEG coherence was used to choose aspects indicative of different brain processes. In the first trial, patients with bipolar disorder and major depressive illness were treated using the improved ant colony optimization algorithm. Subsequently, a multitude of SVM models were developed utilizing the previously specified features. The researchers employed a machine learning approach known as improved ant colony optimization to choose features. This technique draws inspiration from the social behaviors exhibited by natural insects and animals. Once a collection of features had been determined, they were included in the SVM for utilization in the process of pattern identification. The researchers examined the application of SVMs along with four distinct feature selection approaches, one of which was SVM-improved ant colony optimization. A comparative analysis of the performance of SVM-improved ant colony optimization and SVM was done separately. Among the several models evaluated, SVM-improved ant colony optimization exhibited the greatest accuracy rate of 80.19% in discerning the distinction between bipolar disorder and major depressive disorder. This evaluation encompassed a comprehensive set of 22 criteria. Alternative models have 62.37% to 78.21% accuracy with 25 to 48 characteristics [25]. Erguzel et al. [25] employed Cordance's quantitative electroencephalography to distinguish unipolar and bipolar depression. To improve the model without selecting features, the authors used the particle swarm optimization method for feature selection. They then used the chosen EEG alpha and theta frequencies to build an artificial neural network (ANN) model [26]. The study underlines the importance of feature selection for model improvement since it eliminates confusing aspects [25]. Insect swarm optimization (improved ant colony optimization) and population-based swarm optimization represent two instances of animal-inspired algorithms.

7.4.1.2 NEUROPSYCHOLOGICAL TESTS

The investigator developed a cognitive test for the diagnosis of bipolar disorder that utilizes video-based technology and does not require invasive procedures. The research involved the establishment of a system to detect and monitor the pupils of an individual's eyes. Additionally, the device recorded the duration of participants' gaze in various directions and their contemplative activities. The researcher developed a SVM algorithm using data obtained from the pupils of subjects in both the training and test sets. The results of the study indicated that the algorithm effectively distinguished individuals with bipolar disorder from those without (HC) with a high accuracy rate of 96.36% [27].

7.4.1.3 STRUCTURAL NEUROIMAGING

A total of 11 studies have been identified that utilized structural MRI to distinguish individuals with bipolar disorder from those with certain mental health conditions or healthy individuals. Numerous studies have specifically examined the characteristics of white matter and gray matter to ascertain their distinctive features. Mwangi et al. utilized the relevance vector machine algorithm to discern between individuals with bipolar illness and healthy control participants. This discrimination was based on the analysis of gray matter and white matter density in a substantial cohort of 256 patients. The study revealed that the utilization of white matter alone resulted in accuracies of 70.3%, whereas the use of gray matter alone yielded accuracies of 64.9%. When both approaches were employed simultaneously, the accuracy dropped somewhat to 64.4%. This difference in accuracy was statistically significant ($p < 0.005$). The examination of relevance vector machine-predicted probability scores for the three stages of bipolar disorder revealed noteworthy results. Specifically, early-stage bipolar disorder and a healthy control group were found to be statistically indistinguishable ($p = 0.05$). On the other hand, intermediate-stage bipolar disorder and late-stage bipolar disorder exhibited significant differences compared to the healthy control group ($p = 0.01$ and $p = 0.02$, respectively). This strengthens the increasing number of data linking bipolar disorder to a degenerative neurological condition [28]. In an additional investigation, the researchers employed SVM and Gaussian process classification techniques to analyze whole-brain gray matter data from two distinct populations. The objective was to classify individuals into three groups: those with bipolar disorder, those with unipolar depression, and a healthy control group. The authors achieved a classification accuracy of 75.9% (SVM, $p < 0.001$) and 79.3% (Gaussian process classification, $p < 0.001$) in the first sample, and 65.5% (SVM, $p = 0.006$) and 65.5% (Gaussian process classification, $p = 0.006$) in the second sample, effectively distinguishing between bipolar disorder and unipolar depression. There were no observed improvements in accuracy when the white matter was incorporated into the model, similar to the findings reported by Mwangi et al. [28].

7.4.2 APPROACHES FOR SCHIZOPHRENIA PREDICTION

7.4.2.1 SUPPORT VECTOR MACHINE (SVM)

The support vector machine (SVM) approach is commonly utilized in the context of nonlinear input data due to its capacity to leverage kernel functions. Kernel functions include linear kernel, Gaussian radial basis, and polynomial functions. SVMs have been utilized in several studies to diagnose schizophrenia. The structural MRI data have been used to explore the SVM approach [29]. The investigation involved a comparative analysis of voxel-based morphometry to assess the estimated gray matter densities of a sample size of 212 individuals from both the schizophrenia and healthy control groups. To enhance the precision of the results, a cross-validation technique was implemented to validate the findings. The trained data yielded an accuracy rate of 86% from a sample size of 127 individuals, whereas the validation data exhibited an accuracy rate of 83% from a sample size of 85 individuals. The capacity of a SVM to categorize MRI data to differentiate individuals with seizures from those without showcases the efficacy of machine learning techniques in elucidating nonlinear associations between input and output data. This exemplifies the capacity of the model to effectively handle data with many dimensions and mitigate the issue of overfitting.

7.4.2.2 NATURAL LANGUAGE PROCESSING

Natural language processing is a computational technique used to derive semantic understanding from textual input, to train an intelligent system to accurately detect instances of schizophrenia. Natural language processing uses the linguistic context of words to identify pertinent keywords and phrases within a given text. For example, this method depends on the utilization of patients' self-reports to ascertain the particular manifestations and indications they experienced in the context of schizophrenia. Patients cannot be classified based on a single term. Consequently, it is imperative to utilize techniques for obtaining semantic information to discern distinctions among these individuals. In recent times, machine learning algorithms have been employed in the field of natural language processing to enhance learning processes and achieve more efficient results. Several potential sources of textual data are utilized to diagnose schizophrenia. Natural language processing technology can be utilized to diagnose diseases by extracting and analyzing the vast amount of data present on social media platforms [30].

We can forecast the start of psychosis using natural language processing and data from 40 first-episode psychosis interview transcripts. The DNN achieved a classification accuracy of 99% in distinguishing between speech samples from patients and those from healthy individuals. This demonstrates the potential of machine learning in combination with natural language processing techniques for extracting knowledge from spoken language.

7.4.2.3 DEEP NEURAL NETWORK (DNN)

In the context of schizophrenia, an ANN is a very efficacious machine learning methodology for problem resolution. Similar to the human brain, an ANN possesses several synapses, which are connections linking its diverse processing nodes. An ANN consists of three primary layers: the input layer, the hidden layer, and the output layer. The artificial neuron is a mathematical unit for nonlinear transformation. It uses different activation functions, like sigmoid, rectified linear unit, and hyperbolic tangent, to calculate the weighted sum of the nodes in the input layer. The feedforward neural network is the fundamental structure of an ANN. The inclusion of several hidden layers in a neural network contributes to its complexity, rendering it a deep ANN. These layers reveal the presence of nonlinear relationships between the input and output data. DNNs have been utilized in various image processing applications, including the prediction of mental illness through the analysis of MRI images. Convolutional DNNs have been widely employed in mental health disorder identification [31–34].

7.4.2.4 LOGISTIC REGRESSION (LR)

The logistic regression (LR) algorithm is frequently employed to classify problems into two distinct groups. The objective of binary classification is to ascertain the presence or absence of schizophrenia in a patient. LR is a statistical method used to estimate the probability of an event occurring. It utilizes the sigmoid function, as in the case of determining the likelihood of a patient being diagnosed with schizophrenia. Several studies have employed LR as a statistical method for the identification and detection of sarcoidosis. LR has been utilized to identify significant factors in the recovery stages of individuals with schizophrenia within a cohort of 75 participants from Hong Kong [35]. Data collection involved tracking socio-demographic factors, stages of recovery, and various aspects connected to the healing process. The categorization accuracy of LR for stages 3 ("living with disability") and 4 ("living beyond disability") recovery was found to be 75.45% and 75.50%, respectively. LR analysis reveals that age plays a key role in defining the various phases of the healing process.

7.4.2.5 K-FOLD CROSS VALIDATION

The occurrence of overfitting and underfitting can be minimized by employing k-fold cross-validation. In this context, the dataset is divided into *K* distinct categories. In this approach, the model undergoes initial training on a set of K1 categories, followed by evaluation on a set of (K1) + 2 categories. This is repeated *K* times. A median is determined from the outcomes of all validation sets to establish the model's ultimate performance. To achieve generalization of findings, machine learning algorithms utilize k-fold cross-validation. One study trained several classification models for error propagation using five-fold cross-validation [36]. The dataset was divided into two subsets, namely a training set and a validation set, with each subset comprising 80% of the total data. The evaluation metrics, namely F1-score, precision, recall, and accuracy, are subsequently computed on the validation set.

7.4.3 DIAGNOSTIC METHODS FOR POST-TRAUMATIC STRESS DISORDER

Reece et al. analyze Twitter postings to determine if a user would develop major depressive disorder or post-traumatic stress disorder using the machine learning method of random forest (RF) [37]. The researchers thoroughly examined a dataset consisting of over 2,43,000 tweets posted by individuals diagnosed with post-traumatic stress disorder to obtain their results. The authors used RF to predict post-traumatic stress disorder with an area under the curve of 0.89. In their study, Leightley et al. [38] employed machine learning methodologies to forecast the occurrence of post-traumatic stress disorder within the British military services. The researchers utilized a dataset of 13,690 individuals who had served in the armed forces between the years 2004 and 2009. The forecast utilizes many machine learning algorithms. The experimental findings demonstrate that the utilization of RFs yields the highest degree of predictive accuracy, reaching an impressive 97% [38]. Among the various machine learning techniques, it has been shown that the ANN exhibits the lowest level of accuracy, namely at 89% [39]. In contrast, Bagging demonstrates a higher accuracy rate of 95%, while the SVM obtains an accuracy level of 91%. Conrad et al. [40] examined the utilization of machine learning algorithms to forecast post-traumatic stress disorder among Ugandan civil war veterans. The authors employed a training data set including 441 individuals who experienced trauma, whereas a testing data set of 211 individuals was utilized. A diverse range of machine learning techniques, such as RF with conditional inference, least absolute shrinkage and selection, and LR, is being employed to make predictions about individuals who have experienced post-traumatic stress disorder. It was found that the RF with conditional inference has the highest level of reliability [40], compared to the least absolute shrinkage and selection (74.88% accuracy) and LR (75.36% reliability) methods. Marmar et al. used machine learning techniques to predict post-traumatic stress disorder using audio recordings. To gain insight into the communication patterns of individuals who have experienced warfare, the authors collated transcriptions of interviews conducted with military veterans [41]. Clinical interviews also generate speech characteristics that have the potential to serve as predictive indicators of post-traumatic stress disorder, including speech characterized by reduced speed and a more monotonous tone. The prediction model employed in this study is the RF algorithm, which achieved an area under the curve value of 0.954 and an accuracy rate of 89.1% [42].

7.4.4 APPROACHES FOR DEPRESSION AND ANXIETY DETECTION

The field of mental health research faces significant challenges in predicting anxiety due to its clinical parallels with severe depressive disorder [43]. Sau et al. used machine learning approaches to predict depression and anxiety in the elderly [44]. A total of 10 alternative classifiers were assessed using a limited selection of attributes, and it was determined that the RF algorithm achieved the highest level of accuracy, reaching 89%. In another study, Sua et al. employed the Hospital Anxiety and Depression Scale as a predictive tool for assessing anxiety and depression among individuals working in the maritime industry [45]. A comprehensive evaluation was conducted on a set of five machine-learning classifiers. RF achieved 81.2% accuracy and 81.2% precision, respectively, whereas CatBoost achieved 82.6% accuracy and 84.1% precision. Cho et al. introduced the RF method for detecting depression. The system utilizes data obtained from individuals who took part in the medical checkup program of the National Health Insurance Sharing Service of Korea [43]. The study revealed that 0.02% of the participants exhibited clinical depression, while the overwhelming majority of 99.8% did not. Consequently, the researchers were encouraged to contemplate the implementation of down- or up-sampling techniques to establish statistical parity between the two groups. At the end of the study, an area under the curve of 0.849 was recorded. In their paper, Sharma et al. proposed a machine-learning approach using the Lifelines Database, which incorporates self-reported depression data and biomarker data, to enhance the detection of depression [46]. Given the presence of skewed information in the dataset used for this analysis, researchers implemented various resampling techniques to address this issue. The samples were run through an XGBoost (XGB) algorithm (a form of extreme gradient boosting). Supervised learning has been extensively used to predict mental illness [47, 48]. To collect and process textual data, the researchers employed emotional AI that was equipped with classifiers such as naive bayes (NB) and SVMs. The multinomial NB classifier exhibited superior performance compared to the SVM classifier. Hilbert et al. employed supervised learning techniques, namely utilizing a SVM, to analyze multimodal biobehavioral data to differentiate those exhibiting symptoms of anxiety from those experiencing depression [49]. This study used information from clinical questionnaires, cortisol levels, structural brain measures, and gray and white matter quantities. The study concluded that the clinical questionnaires are inadequate for anxiety classification. However, using cortisol and gray matter volume data helped with anxiety classification. Using cognitive behavioral performance data coupled with machine learning, Richter et al. suggested a novel and objective diagnostic technique for distinguishing anxiety from depression [48]. Participants in the subclinical range who also had high levels of depression and anxiety were given questionnaires to fill out. The impact of their biases on their cognitive processes was assessed by the administration of six distinct cognitive-behavioral tasks. The RF algorithm was employed to systematically allocate individuals based on their comprehensive learning and memory performance, following the collection and preparation of the data.

7.4.5 APPROACHES FOR ATTENTION DEFICIT HYPERACTIVITY DISORDER (ADHD) DETECTION

Mikolas et al. [50] presented a methodology for distinguishing individuals diagnosed with attention deficit hyperactivity disorder (ADHD) from those with other mental disorders by utilizing de-identified clinical information. In this study, the SVM classifier included a total of 30 features. Additionally, a secondary classification approach was utilized, which did not consider the demographic attributes of the participants, such as gender or age. Furthermore, a secondary classification method was employed to handle missing data. The accuracy rates of the two entities were 65.1% and 68.8%, respectively, indicating commendable performance [50]. Tan et al.

blinded group-level MRI imaging [51]. Functional volumes of the brain were computed using the fMRI data. The predicted regional brain volumes from imaging data were compared. Overall, 67% of ADHD patients were properly categorized by SVM classifiers that had been trained on functional volumes and demographic data. Like children, adults have the potential to develop ADHD. Batsakis et al. examined trends in adult ADHD [52]. The mixed ML-KBM model was utilized to analyze both clinical data and survey responses. The effectiveness of this strategy in clinical trials is 95% accurate. Peng et al. [53] presented an AI-based diagnostic tool for ADHD. The evaluation of features from a dataset consisting of 110 individuals was conducted using extreme learning machine and SVM algorithms, employing the leave-one-out cross-validation technique. In comparison to the SVM, the extreme learning machine exhibited a higher accuracy rate of 90.18%, while the SVM had a lower accuracy rate of 86.55% [53].

7.5 CONCLUSION

A wide range of approaches and algorithms are accessible for the diagnosis and treatment of mental health conditions. Numerous existing methodologies exhibit potential for enhancement. The area of machine learning for mental health is still in its nascent stages, with significant room for advancement in terms of identifying novel challenges and testing potential solutions across diverse settings. The task of categorizing mental health data has inherent challenges, and the selection of features included in machine learning algorithms plays a crucial role in determining the effectiveness of classification outcomes. Consequently, a significant portion of research and studies continues to encounter challenges in substantiating their findings, primarily attributable to a dearth of credible and validated information, particularly from external sources. Moreover, the efficacy of most machine learning methodologies exhibits variability contingent upon the nature of the problem at hand. The accuracy of machine learning models is influenced by the quality of the data used for training. However, the outcomes of machine learning models can be influenced by pre-processing techniques such as data cleansing and parameter optimization. To determine the most optimal machine learning algorithm, researchers must conduct thorough testing and analysis of the data using several machine learning algorithms.

KEYWORDS

- **artificial intelligence**
- **attention deficit hyperactivity disorder**
- **functional volumes**
- **machine learning algorithms**
- **naive bayes**
- **secondary classification approach**

REFERENCES

1. World Health Organization. (2004). *Promoting Mental Health: Concepts, Emerging Evidence, Practice: Summary Report.* World Health Organization.
2. Abd Rahman, R., Omar, K., Noah, S. A., & Danuri, M. M. (2018). A survey on mental health detection in online social networks. *International Journal on Advanced Science, Engineering and Information Technology*, *8*(4–2), 1431.
3. Vos, T., Barber, R. M., Bell, B., Bertozzi-Villa, A., Biryukov, S., Bolliger, I., Charlson, F., Davis, A., Degenhardt, L., Dicker, D., & Duan, L. (2015). Global, regional, and national incidence, prevalence, and years lived with disability for 301 acute and chronic diseases and injuries in 188 countries, 1990–2013: A systematic analysis for the Global Burden of Disease Study 2013. *The Lancet*, *386*(9995), 743–800.
4. Divsalar, D., Jin, H., & McEliece, R. J. (1998). Coding theorems for "turbo-like" codes. In *Proceedings of the annual Allerton Conference on Communication Control and Computing* (Vol. 36, pp. 201–210). University of Illinois.
5. World Health Organization. (2019). *Global Action Plan on Physical Activity 2018–2030: More Active People for a Healthier World.* World Health Organization.
6. Baker, A. J., & Verrocchio, M. C. (2016). Exposure to parental alienation and subsequent anxiety and depression in Italian adults. *The American Journal of Family Therapy*, *44*(5), 255–271.
7. Keyes, C. L. (2007). Promoting and protecting mental health as flourishing: A complementary strategy for improving national mental health. *American Psychologist*, *62*(2), 95.
8. Hwang, S. J. (n.d.). Discriminative object categorization with external semantic knowledge. https://repositories.lib.utexas.edu/items/e77deb6f-0c18-4e6a-b0a9-0d2e5720fd70.
9. Garcia-Ceja, E., Osmani, V., & Mayora, O. (2015). Automatic stress detection in working environments from smartphones' accelerometer data: A first step. *IEEE Journal of Biomedical and Health Informatics*, *20*(4), 1053–1060.

10. Sano, A., & Picard, R. W. (2013). Stress recognition using wearable sensors and mobile phones. In *2013 Humaine Association Conference on Affective Computing and Intelligent Interaction* (pp. 671–676). IEEE.
11. Vrijkotte, T. G., Van Doornen, L. J., & De Geus, E. J. (2000). Effects of work stress on ambulatory blood pressure, heart rate, and heart rate variability. *Hypertension, 35*(4), 880–886.
12. Ali, G., Ali, A., Ali, F., Draz, U., Majeed, F., Yasin, S., Ali, T., & Haider, N. (2020). Artificial neural network-based ensemble approach for multicultural facial expressions analysis. *IEEE Access, 8*, 134950–134963.
13. Dardenne, A., Baeken, C., Crunelle, C. L., Bervoets, C., Matthys, F., & Herremans, S. C. (2018). Accelerated HF-rTMS in the elderly depressed: A feasibility study. *Brain Stimulation: Basic, Translational, and Clinical Research in Neuromodulation, 11*(1), 247–248.
14. Torrey, L., & Shavlik, J. (2010). Transfer learning. In *Handbook of Research on Machine Learning Applications and Trends: Algorithms, Methods, and Techniques* (pp. 242–264). IGI Global.
15. LeCun, Y., Bengio, Y., & Hinton, G. (2015). Deep learning. *Nature, 521*(7553), 436–444.
16. Graham, S., Depp, C., Lee, E. E., Nebeker, C., Tu, X., Kim, H. C., & Jeste, D. V. (2019). Artificial intelligence for mental health and mental illnesses: An overview. *Current Psychiatry Reports, 21*, 1–8.
17. Lautman, Z., & Lev-Ari, S. (2022). The use of smart devices for mental health diagnosis and care. *Journal of Clinical Medicine, 11*(18), 5359.
18. Judd, L. L., Schettler, P. J., & Akiskal, H. S. (2002). The prevalence, clinical relevance, and public health significance of subthreshold depressions. *Psychiatric Clinics, 25*(4), 685–698.
19. McGorry, P. D., Yung, A. R., Phillips, L. J., Yuen, H. P., Francey, S., Cosgrave, E. M., Germano, D., Bravin, J., McDonald, T., Blair, A., & Adlard, S. (2002). Randomized controlled trial of interventions designed to reduce the risk of progression to first-episode psychosis in a clinical sample with subthreshold symptoms. *Archives of General Psychiatry, 59*(10), 921–928.
20. Olfson, M., Kessler, R. C., Berglund, P. A., & Lin, E. (1998). Psychiatric disorder onset and first treatment contact in the United States and Ontario. *American Journal of Psychiatry, 155*(10), 1415–1422.
21. Coryell, W., Turvey, C., Endicott, J., Leon, A. C., Mueller, T., Solomon, D., & Keller, M. (1998). Bipolar I affective disorder: Predictors of outcome after 15 years. *Journal of Affective Disorders, 50*(2, 3), 109–116.
22. Omar, K. S., Mondal, P., Khan, N. S., Rizvi, M. R., & Islam, M. N. (2019). A machine learning approach to predict autism spectrum disorder. In *2019 International Conference on Electrical, Computer and Communication Engineering (ECCE)* (pp. 1–6). IEEE.
23. Khan, K., & Katarya, R. (2023). Machine learning techniques for autism spectrum disorder: Current trends and future directions. In *2023 4th International Conference on Innovative Trends in Information Technology (ICITIIT)* (pp. 1–7). IEEE.
24. Katarya, R., & Maan, S. (2020). Predicting mental health disorders using machine learning for employees in technical and non-technical companies. In *2020 IEEE*

International Conference on Advances and Developments in Electrical and Electronics Engineering (ICADEE) (pp. 1–5). IEEE.

25. Erguzel, T. T., Sayar, G. H., & Tarhan, N. (2016). Artificial intelligence approach to classify unipolar and bipolar depressive disorders. *Neural Computing and Applications*, *27*(6), 1607–1616.
26. Metin, S. Z., Erguzel, T. T., Ertan, G., Salcini, C., Kocarslan, B., Cebi, M., Metin, B., Tanridag, O., & Tarhan, N. (2018). The use of quantitative EEG for differentiating frontotemporal dementia from late-onset bipolar disorder. *Clinical EEG and Neuroscience*, *49*(3), 171–176.
27. Frith, U., & Happé, F. (2005). Autism spectrum disorder. *Current Biology*, *15*(19), R786–R790.
28. Mwangi, B., Wu, M. J., Bauer, I. E., Modi, H., Zeni, C. P., Zunta-Soares, G. B., Hasan, K. M., & Soares, J. C. (2015). Predictive classification of pediatric bipolar disorder using atlas-based diffusion-weighted imaging and support vector machines. *Psychiatry Research: Neuroimaging*, *234*(2), 265–271.
29. Winterburn, J. L., Voineskos, A. N., Devenyi, G. A., Plitman, E., de la Fuente-Sandoval, C., Bhagwat, N., Graff-Guerrero, A., Knight, J., & Chakravarty, M. M. (2019). Can we accurately classify schizophrenia patients from healthy controls using magnetic resonance imaging and machine learning? A multi-method and multi-dataset study. *Schizophrenia Research*, *214*, 3–10.
30. Vieira, S., Pinaya, W. H., & Mechelli, A. (2017). Using deep learning to investigate the neuroimaging correlates of psychiatric and neurological disorders: Methods and applications. *Neuroscience & Biobehavioral Reviews*, *74*, 58–75.
31. Riaz, A., Asad, M., Al Arif, S. M., Alonso, E., Dima, D., Corr, P., & Slabaugh, G. (2018). Deep fMRI: An end-to-end deep network for classification of fMRI data. In *2018 IEEE 15th International Symposium on Biomedical Imaging (ISBI 2018)* (pp. 1419–1422). IEEE.
32. Zou, L., Zheng, J., Miao, C., Mckeown, M. J., & Wang, Z. J. (2017). 3D CNN based automatic diagnosis of attention deficit hyperactivity disorder using functional and structural MRI. *IEEE Access*, *5*, 23626–23636.
33. Geng, X. F., & Xu, J. H. (2017). Application of autoencoder in depression diagnosis. *DEStech Trans Comput Sci Eng (CSMA)*, 146–151.
34. Borgwardt, S., Koutsouleris, N., Aston, J., Studerus, E., Smieskova, R., Riecher-Rössler, A., & Meisenzahl, E. M. (2013). Distinguishing prodromal from first-episode psychosis using neuroanatomical single-subject pattern recognition. *Schizophrenia Bulletin*, *39*(5), 1105–1114.
35. McManus, K., Mallory, E. K., Goldfeder, R. L., Haynes, W. A., & Tatum, J. D. (2015). Mining Twitter data to improve detection of schizophrenia. *AMIA Summits on Translational Science Proceedings*, 122.
36. Ge, F., Li, Y., Yuan, M., Zhang, J., & Zhang, W. (2020). Identifying predictors of probable posttraumatic stress disorder in children and adolescents with earthquake exposure: A longitudinal study using a machine learning approach. *Journal of Affective Disorders*, *264*, 483–493.
37. Reece, A. G., Reagan, A. J., Lix, K. L., Dodds, P. S., Danforth, C. M., & Langer, E. J. (2017). Forecasting the onset and course of mental illness with Twitter data. *Scientific Reports*, *7*(1), 13006.

38. Leightley, D., Williamson, V., Darby, J., & Fear, N. T. (2019). Identifying probable post-traumatic stress disorder: Applying supervised machine learning to data from a UK military cohort. *Journal of Mental Health, 28*(1), 34–41.
39. Ayala Solares, J. R., Canoy, D., Raimondi, F. E., Zhu, Y., Hassaine, A., Salimi-Khorshidi, G., Tran, J., Copland, E., Zottoli, M., Pinho-Gomes, A. C., & Nazarzadeh, M. (2019). Long-term exposure to elevated systolic blood pressure in predicting incident cardiovascular disease: Evidence from large-scale routine electronic health records. *Journal of the American Heart Association, 8*(12), e012129.
40. Conrad, D., Wilker, S., Pfeiffer, A., Lingenfelder, B., Ebalu, T., Lanzinger, H., Elbert, T., Kolassa, I. T., & Kolassa, S. (2017). Does trauma event type matter in the assessment of traumatic load? *European Journal of Psychotraumatology, 8*(1), 1344079.
41. Marmar, C. R., Brown, A. D., Qian, M., Laska, E., Siegel, C., Li, M., Abu-Amara, D., Tsiartas, A., Richey, C., Smith, J., & Knoth, B. (2019). Speech-based markers for posttraumatic stress disorder in US veterans. *Depression and Anxiety, 36*(7), 607–616.
42. Aleem, S., Huda, N. U., Amin, R., Khalid, S., Alshamrani, S. S., & Alshehri, A. (2022). Machine learning algorithms for depression: Diagnosis, insights, and research directions. *Electronics, 11*(7), 1111.
43. Cho, S. E., Geem, Z. W., & Na, K. S. (2020). Prediction of depression among medical check-ups of 433,190 patients: A nationwide population-based study. *Psychiatry Research, 293*, 113474.
44. Sau, A., & Bhakta, I. (2019). Screening of anxiety and depression among seafarers using machine learning technology. *Informatics in Medicine Unlocked, 16*, 100228.
45. Sau, A., & Bhakta, I. (2017). Predicting anxiety and depression in elderly patients using machine learning technology. *Healthcare Technology Letters, 4*(6), 238–243.
46. Sharma, A., & Verbeke, W. J. (2021). Understanding the importance of clinical biomarkers for diagnosis of anxiety disorders using machine learning models. *PloS One, 16*(5), e0251365.
47. Deshpande, M., & Rao, V. (2017). Depression detection using emotion artificial intelligence. In *2017 International Conference on Intelligent Sustainable Systems (ICISS)* (pp. 858–862). IEEE.
48. Richter, T., Fishbain, B., Fruchter, E., Richter-Levin, G., & Okon-Singer, H. (2021). Machine learning-based diagnosis support system for differentiating between clinical anxiety and depression disorders. *Journal of Psychiatric Research, 141*, 199–205.
49. Hilbert, K., Lueken, U., Muehlhan, M., & Beesdo-Baum, K. (2017). Separating generalized anxiety disorder from major depression using clinical, hormonal, and structural MRI data: A multimodal machine learning study. *Brain and Behavior, 7*(3), e00633.
50. Mikolas, P., Vahid, A., Bernardoni, F., Süß, M., Martini, J., Beste, C., & Bluschke, A. (2022). Training a machine learning classifier to identify ADHD based on real-world clinical data from medical records. *Scientific Reports, 12*(1), 12934.
51. Tan, W., Xu, Y., Liu, P., Liu, C., Li, Y., Du, Y., Chen, C., Wang, Y., & Zhang, Y. (2021). A method of VR-EEG scene cognitive rehabilitation training. *Health Information Science and Systems, 9*, 1–9.

52. Batsakis, S., Papadakis, E., Tachmazidis, I., Chen, T., Antoniou, G., & Adamou, M. (2023). Neuro Intel: A system for clinical diagnosis of attention deficit hyperactivity disorder (ADHD) using artificial intelligence. In *2023 IEEE Symposium on Computers and Communications (ISCC)* (pp. 1–6). IEEE.
53. Peng, X., Lin, P., Zhang, T., & Wang, J. (2013). Extreme learning machine-based classification of ADHD using brain structural MRI data. *PloS One*, *8*(11), e79476.

CHAPTER 8

Machine Learning Approaches for Electronic Health Record Phenotyping

ABSTRACT

The rising utilization of electronic health information in medical research has amplified the demand for accurate and efficient phenotyping methods. Early phenotyping efforts relied on rule-based algorithms that necessitated manual editing by experts. However, in the past few years, machine learning technologies have emerged as a replacement, enabling enhanced scalability across various phenotypes and healthcare environments. This chapter extensively discussed deep learning (DL) models for electronic health record (EHR) data and investigated the potential applications of various deep neural networks (DNNs) for analyzing diverse data sources and fulfilling specific objectives.

8.1 INTRODUCTION

The use of electronic health records (EHRs) is critical in the field of human anatomy research [1]. In recent years, there has been an increase in the use of EHR data for identifying disease genomes, accelerating, and broadening the recruitment process for clinical trials, and advancing epidemiological studies about previously unknown and recently discovered diseases [2–6]. The use of EHRs offers promise as a vital data reservoir for the innovation and implementation of individualized treatment plans, as well as the generation of empirical evidence in real-world settings [7, 8]. Phenotyping, or the identification of patients with specific phenotypes based on information contained in their EHRs, is a critical component for any secondary use of EHR data [9–11]. Phenotyping is an important first

Machine Learning in Healthcare: Advances and Future Prospects. Rishabha Malviya, Niranjan Kaushik, Tamanna Rai, M. P. Saraswathy, and Rajendra Awasthi (Authors)

step in any EHR-based application since it aids in the identification and definition of the population under study.

Phenotyping typically requires four separate stages: data preparation, algorithmic creation, algorithm evaluation, and algorithm implementation (Figure 8.1). The first stage entails the integration of data presentation, which comprises two primary phases. The initial stage is extracting and analyzing relevant data from the data warehouse containing candidate patient records. The second phase involves the manual inspection of a subset of charts to generate phenotypic designations that serve as the gold standard. In step two, researchers create phenotyping algorithms using the data gathered in stage one, which is often known as the data mart. These algorithms are built using either rule-based or machine-learning methodologies. During the third stage, the algorithm's phenotyping is evaluated by comparing it to the gold standard label. Various accuracy metrics, such as positive predictive value, sensitivity, and other measures, are used to analyze the algorithm's performance. Stage four utilizes the approach created in stage two to form a cohort of individuals that have similar phenotypic features. This cohort makes it easier to conduct further investigations. This chapter presents a concise overview of the most recent progress in deep learning (DL) models utilized for analyzing EHR data. Furthermore, it delves into prospective areas for further investigation in this domain.

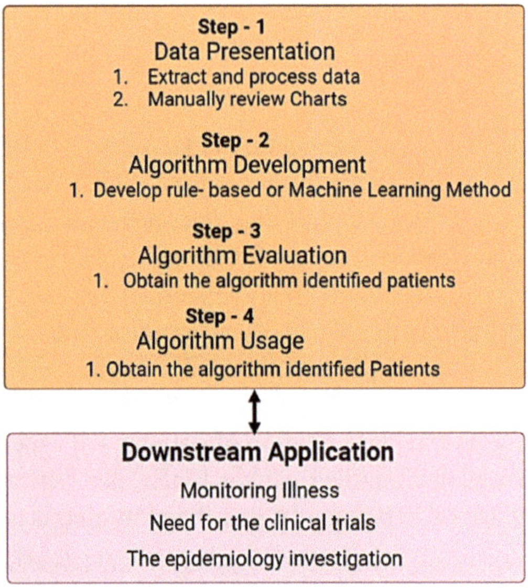

FIGURE 8.1 Presentation of the phenotyping process.

Conventional health analytics models may necessitate time-consuming methods like expert-defined phenotyping and ad hoc feature engineering. Hence, the generated models may not be suitable for exploitation by other entities or datasets. DL has made significant progress in multiple domains, such as speech recognition, machine vision, image classification, and natural language processing. The current shift in data analytic modeling represents a significant transition from the conventional method of expert-driven feature engineering to the modern approach of data-driven feature construction. Recently, there has been an increasing amount of scholarly research that confirms the efficacy of feature building using DL methods. The application of DL in healthcare has attracted unprecedented attention. DL models have exhibited improved performance in comparison to conventional machine learning methods and possess the benefit of necessitating less manual feature engineering. Therefore, these models are extremely appropriate for implementation in healthcare research. In addition, the healthcare industry provides a wide range of complex and detailed datasets that are very suitable for training advanced DL models. Conversely, EHR data poses many unique challenges for DL investigations.

8.2 ELECTRONIC HEALTH RECORD (EHR)

Electronic health record (EHR) management and maintenance in data-driven medicine is seen as a very promising step forward in the healthcare field because it can use large amounts of medical data to make treatment plans that are specific to each patient. The use of EHR is essential to promoting the progress of data-driven healthcare. When using an EHR, individuals may face difficulties such as timeliness, reparability, audibility, bias, and similar concerns. Hence, it is crucial to obtain accurate phenotyping or feature extraction from patients' EHR as a key prerequisite for the successful implementation of any subsequent applications.

A large body of literature focused on data analytics related to patients with EHR has been published in the past 2–3 decades. Wang et al. explored the application of multilinear sparsity LR for forecasting the risk levels of patients based on their EHRs [12]. Zhang et al. introduced a system that relies on similarity to generate individualized therapy recommendations. A composite distance-metric learning technique has been suggested to compare patients from different organizations while keeping personal information safe [13]. The process of extracting significant data from patient EHRs, known as electronic phenotyping in the field of medical informatics, is essential for the advancement of medical applications [14–16]. Although numerous computational models have been developed for electronic phenotyping using EHRs, there are still several obstacles that remain unresolved, including the utilization of a tensor-based approach [17, 18]. Modifications are implemented in the patients' EHRs over time. Valuable insights into impending patient health conditions can be derived from the order in which health issues occur. Due to the intricate nature of patient disorders, it is not uncommon to observe variations in EHRs among individuals, even when they have the same disease. Clinical research that relies exclusively on healthcare records may be significantly biased due to the aforementioned issues as well as inherent flaws within the system [19].

8.3 COMPONENTS OF HEALTH OUTCOMES

Machine learning is particularly valuable for computational phenotyping in the context of categorizing health outcomes in four different scenarios. It is crucial to examine the conventional procedure of recording diagnostic data on health outcomes in EHR. Figure 8.2 presents an illustration of common instances of health outcomes and their correlation with the significance of machine learning findings obtained from EHR data.

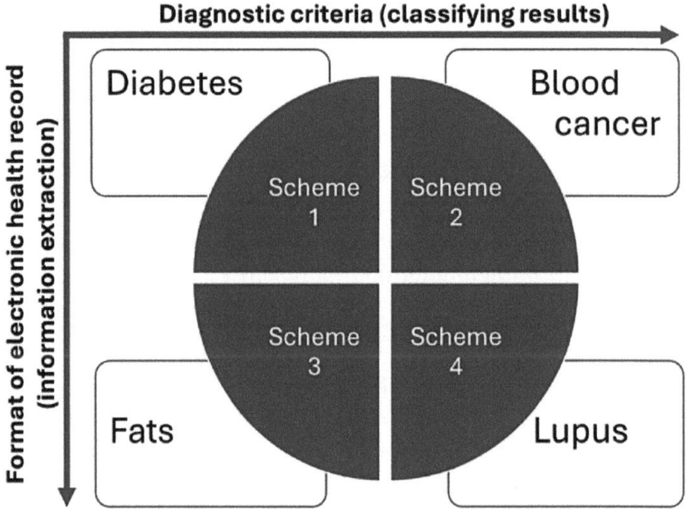

FIGURE 8.2 Illustration of common instances of health outcomes and their correlation with the significance of machine learning findings obtained from electronic health record data.

8.4 ELECTRONIC PHENOTYPING

Electronic health records (EHRs) electronic phenotyping involves extracting and identifying phenotypic information from electronic health data. In other words, electronic phenotyping is the term used to describe the process of obtaining useful phenotypes from longitudinal patient EHRs. Medical research, clinical care, and population health management depend on phenotyping, which characterizes observable traits or clinical features of individuals. Through the utilization of the vast amount of data that is stored in the EHR, electronic phenotyping can automate and simplify the process of identifying particular phenotypes or clinical problems. Conducting this step is necessary prior to implementing any data-driven applications involving EHR, such as predictive modeling and similar tasks [20, 21].

8.4.1 DATA REPRESENTATION BASED ON VECTORS

This method generates a vector that encompasses information pertaining to each individual patient. The dimensionality of each dimension corresponds to the number of distinct medical events in the EHR. Each dimension provides a statistical summary, such as the sum, average, maximum, minimum, and so on, of the linked medical event within the defined time period. The medical events that serve as the basis are considered vectors, and the phenotypes are determined by combining these vectors using coefficients derived through optimization techniques [22]. This approach fails to account for temporal relationships between events.

8.4.2 DATA REPRESENTATION BASED ON SEQUENCE

By going through this process, a chronological order of EHRs is generated for each individual patient. Pattern mining techniques can be used once the temporal patterns have been recognized as attributes [23, 24]. The diversity of patient EHRs often leads to a significant proliferation of patterns, commonly referred to as pattern explosion. Evaluating the therapeutic usefulness of a phenotype is a difficult undertaking.

8.4.3 REPRESENTATION IN TERMS OF TENSORS

This method is used to generate a tensor of the EHR for each patient. Each tensor mode corresponds to a unique medical entity. The value of each slot is a statistical aggregation of all potential event coincidences across the relevant dimensions. Ho et al. suggested employing a nonnegative tensor factorization-based technique to extract phenotypes from these EHR tensors. This approach enabled us to examine the interaction among multiple medical components. However, they fail to consider the simultaneous occurrence of events over time [25].

8.4.4 EXTENDING MATRIX REPRESENTATION THEORY TO TIME

This approach involves representing EHRs as temporal matrices, which consist of two dimensions: time and medical events. Zhou et al. proposed a phenotyping technique that involves clustering medical events with similar temporal patterns. However, they failed to consider the interconnectedness of events [26]. Wang et al. developed a convolutional matrix factorization method to identify shift-invariant patterns in patient EHR matrices. However, they faced the challenge of determining the optimal feature lengths and had to resort to enumerating all potential combinations instead [19].

8.5 ELECTRONIC HEALTH RECORD (EHR) DATA FORMAT

Electronic health records (HERs) systems have the capability to record clinical data generated during the testing process in both structured and unstructured formats. The EHR stores coded billing data, such as diagnostic and treatment codes, laboratory test results, and vital signs. This allows for easy access to a wide range of clinical information. The composition of EHR data consists of 80% unstructured text and images, encompassing admission and discharge summaries, progress notes, pathology, and radiology test findings, and other relevant information [27]. Structured EHR data cannot be retrieved using the same procedures, and conducting a large-scale human review is not feasible [28]. Machine learning can be used to extract and organize important clinical information from unstructured EHR data for tasks such as natural language processing or picture identification. Computer speech recognition and natural language processing applications are experiencing substantial advancements due to the implementation of DL techniques [29].

8.6 ANALYSIS OF PROJECTS USING ELECTRONIC HEALTH RECORD (EHR) DATA

8.6.1 CATEGORIZATION OF DISEASE

The process of developing a DL model for disease classification involves using multiple layers of neural networks to establish a connection between the input EHR data and the desired disease outcome. Some sections utilize data sets that are specialized for certain conditions, such as data from the Parkinson's Progression Markers Initiative and the Pooled Resource of Open-Access Amyotrophic Lateral Sclerosis Clinical Trials [30, 31]. Multiple studies utilize multimodal data and can perform either binary classification or multi-class classification [31]. In addition to the multimodal data obtained from individuals with the disease, other research has also utilized multivariate time series data. In a previous study, researchers utilized convolutional neural networks (CNNs) to analyze multivariate encephalogram (EEG) signals. The objective was to automatically classify the signals into three categories: regular, pre-ictal, and seizure [32]. Making use of information from the Medical Information Mart for Intensive Care III (MIMIC III) [33], a long short-term memory (LSTM) model has been developed for the purpose of diagnosing sepsis [34–36]. Automatic diagnosis or disease code classification of clinical records are common examples of multilabel classification challenges [37, 38]. The convolution plus attention model was employed in 2018 to elucidate clinical notes into diagnosis codes [39]. Deep feedforward and CNNs are used to automatically extract the primary cancer site and the direction of the cancer from free text pathology data [40–42].

8.6.2 CONCEPT EMBEDDING

Clinical phenotyping, a specific form of idea embedding, involves assigning phenotypes to EHR data. Phenotypic characteristics can be derived by utilizing broad concept embedding techniques, such as med2vec [43]. DL techniques are frequently employed for training unsupervised idea embeddings. Large EHR databases are commonly utilized to obtain a high level of generalizability. The Mount Sinai data warehouse utilized 7,00,000 digitized health records to create a patient representation [44]. The study investigated several shallow feature learning techniques, including principal component analysis, k-means clustering, and the Gaussian mixture model, to determine the most effective approach for generating an idea embedding. The resulting embedding was then evaluated through disease prediction tasks. Testing with concept embedding yielded better disease prediction results when compared to conventional feature learning methods. Concept embedding models have been applied to a dataset of 5,50,339 patient records from Children's Hospital of Atlanta (CHOA), which led to significant improvements in performance across several real-world prediction tasks [42–44]. Concept embedding has been used to extract specific medical ideas from MIMIC III discharge data and has also been used to predict phenotypes successfully. The study revealed that DL models, such as random forest (RF), performed less effectively than shallow models (e.g., logistic regression (LR)) when the training sample size was limited [45, 46].

8.6.3 SEQUENTIAL PREDICTION OF CLINICAL EVENTS

Neuronal networks have been used to successfully find links between existing data and possible outcomes through the modeling of longitudinal EHR data. Recurrent neural networks (RNN) and Sutter Health longitudinal outpatient data have been reported to predict the onset of heart failure in the emergence of a completely new disease [47]. The deep feedforward neural network had the largest area under the curve when it came to forecasting the next hospital admission using the 1,328,384 patients (3,295,775 visits) from the New Zealand National Minimum Dataset. A total of 1,14,000 patient records from the University of California, San Francisco (2012–2016) and the University of Chicago Medicine (2009–2016) were used for the prediction tasks. Three distinct DL models were tested: (i) based on RNN; (ii) based on recognition and time awareness; and (iii) combining a neural network model with a neural network model that included decision stumps based on boosted time. Without requiring the harmonization of site-specific data, DL algorithms were able to predict in-hospital mortality, recurrence, hospitalization time, and discharge diagnoses with a high degree of predictability across sites [48, 49].

8.6.4 ELECTRONIC HEALTH RECORD (EHR) DATA PRIVACY

Before maintaining confidentiality, the information contained in a patient's EHR must be extracted from its original context. Dernoncourt et al. developed a de-identification method based on RNN using the i2b2 2014 data and the MIMIC de-identification data. RNN demonstrates better performance compared to other currently employed techniques [50]. A hybrid model consisting of an RNN has been developed to de-identify clinical notes. For character-level representation, a bidirectional LSTM model can be employed [51].

8.7 FRAMEWORKS FOR DEEP LEARNING (DL) IN ANALYTICS

Deep learning (DL) can help in obtaining abstract representations of data for multi-layered computational models. The ability to diagnose diabetic retinopathy using deep neural networks (DNNs) is an example of how machine learning has significantly improved outcomes in the medical and healthcare domains. This has also been noted in image processing [52, 53], speech recognition [54], and natural language processing domains [53].

8.7.1 CONVOLUTIONAL NEURAL NETWORK (CNN)

A convolutional neural network (CNN) can extract complex patterns from images, audio, and video by using the characteristics of localized data. CNNs have significantly enhanced the automated classification of skin lesions from images. Convolutional layers can be used to create local features that are translation-independent by linking multiple local filters with their respective input data. With the help of pooling layers, it is possible to prevent overfitting by gradually decreasing the output rate. In the context of image analysis, it is important to note that the expression of a local feature does not have any impact on other regions. Similarly, the operations of convolution and pooling, which are carried out locally, also do not affect other regions. We need to carefully consider how to effectively capture temporal information when using CNNs for modeling because time-series EHR data is so important [55]. A hybrid convolutional RNN would be the most effective method for both the extraction of joint characteristics and the development of a temporal summary. Although CNNs were initially developed to model images and events, they have now discovered other uses, such as the interpretation of medical writing.

8.7.2 UNSUPERVISED EMBEDDING

Convolutional neural networks (CNNs) can extract complex patterns from images, audio, and video by using the characteristics of localized data. CNNs have significantly enhanced the automated classification of skin lesions from images. A CNN works by connecting several local filters to the input data through convolutional layers. This creates local features that do not change when the data is moved. Reducing the output via pooling layers can help avoid overfitting. Local image analysis algorithms like convolution and pooling barely affect nearby regions. When utilizing a CNN to encode temporal EHR data, it is crucial to carefully address the method of capturing temporality, as this type of data often contains valuable information [32, 55]. For extracting characteristics in the form of a vector and simultaneously summarizing them over time, a hybrid convolutional RNN can be utilized. Beyond the boundaries of image and event modeling, CNNs have a wide range of applications, including the recognition of textual formats that are utilized in the healthcare sector [38, 40].

8.7.3 AUTOENCODER

A nonlinear transformation known as the autoencoders model can reduce the number of dimensions without the assistance of a human being doing so. The autoencoder model family is commonly employed for medical concept integration, such as translating across different medical coding systems [56–62]. The inputs are encoded by the autoencoders using a low-dimensional code representation, and the outputs are then decoded back into the input space. The structure formed by combining the encoder and decoder is known as the reconstruction function. In a standard auto-encoder implementation, the number of dimensions is reduced, but the capacity to capture the most critical aspects of the data is retained. Using autoencoders for unsupervised modeling is a good way to represent EHRs with long-lasting structures and recurring patterns. Sparse artificial evolution can be categorized into two categories: (i) demonizing autoencoders; and (ii) low-frequency autoencoders (SAE). Sparse representation can be obtained by incorporating a sparsity penalty and SAE enabling into the internal code representation, regularizing reconstruction loss. The application of SAE is beneficial in different settings, such as the unsupervised phenotyping of EHRs and the sparse representation of EEG data [63, 64]. The denoising autoencoder (DAE) model exhibits resistance to missing or noisy data due to its construction using purposely distorted inputs. DAE has been employed to generate precise models of human physiology, extract meaningful phenotypes from EHR, and identify correlations between diseases and genes.

8.7.4 RECURRENT NEURAL NETWORKS (RNNS)

Recurrent neural networks (RNNs) are neural networks that can extend feedforward models to simulate time series, event sequences, and natural language text. RNNs are the ideal HER modeling architecture [65–67] for a range of applications, such as the prediction of sequential clinical occurrences and the categorization of diseases [68–73]. RNNs are used in computational phenotyping to capture the intricate temporal dynamics of longitudinal EHR data. RNNs are the most effective architecture for EHR modeling [74]. The hidden layer of the RNN encompasses the memory of the network, which is dependent upon the input and the state of the network. RNNs can process sequences of various lengths. LSTM and gated recurrent units are two well-known examples of gated RNN architectures. They intend to address the issue of vanishing gradients and the problem of long-term reliance.

8.7.5 GENERATIVE ADVERSARIAL NETWORK

A generative adversarial network is a type of machine learning model that specifically deals with the production of data using a game-theoretic process. To accomplish this, a neural network consisting of a generator and a discriminator is trained. The discriminator possesses the ability to differentiate between authentic and counterfeit samples, while the generator can produce samples in response to random input. Both networks are trained sequentially to enhance the generator's output and the discriminator's ability to distinguish between real and fake samples. Recently, the healthcare industry has employed generative adversarial networks to produce discrete codes and continuous medical time series.

8.8 SPECIFIC DIFFICULTIES AND POTENTIAL APPROACHES

The unique challenges posed by EHR data include temporal irregularity, nature, multimodality, lack of labeling, and the complexity of the models themselves. These factors make it difficult to interpret the data accurately [75].

8.8.1 TEMPORALITY AND IRREGULARITY

Longitudinal EHR data provides valuable insights into the progression of patients' health conditions over a period. Long-term effects provide the global context, and doctors use the local context to build a patient's medical history through the connections between the medical events recorded in EHRs. Clinical variables, including diagnostic tests, surgeries, and medications, have an impact on patient health outcomes. The environment, such as the likelihood of readmission or the presence of a disease, also has an impact on these health outcomes. However, due to the complex relationships between different clinical events, it might be challenging to distinguish actual indications over a period. Several researchers have identified significant disparities in the quantity of data present within individual patient records when events were selected randomly. If this anomaly is not addressed appropriately, it would negatively impact the model's efficacy.

8.8.2 LACK OF LABELS

Labels pertain to the tangible conditions of clinical results or the characteristics of diseases in their physical appearances. The absence of specific gold standard labels in EHR data is a frequent occurrence, posing a challenge to training models with enough labels. A major obstacle to utilizing DL for EHR data is developing efficient techniques for classifying EHR information. Generating labels is a proficiency that usually requires the assistance of experienced experts in the specific field who are knowledgeable in the presentation. In practice, the term "silver standard" is commonly employed. For instance, a survey developed patient labels by analyzing the frequency of codes such as treatment, analysis, and dosage regimen in many articles that employed supervised learning methodology. These codes were found in most of the publications analyzed. Transfer learning can be employed as an alternative to manually created labels to avoid their usage. LSTM can reproduce sequences of diagnostic codes, and it can be applied to many datasets with the same level of success. To forecast an outcome, a customized autoencoder structure can be utilized to transfer knowledge from a generic EHR. The prediction of prescriptions can only be based on diagnostics [43].

8.8.3 MULTIMODALITY

Electronic health record (EHR) contains prognostic information, drug, and therapy codes, continuous monitoring data from ECG and EEG devices, as well as medical photographs. The ability to identify patterns in multimodal data is the outcome of effectively diagnosing a problem, making precise predictions, and attaining resilient performance in a learning system [76]. However, the inherent variability of the data makes multimodal learning difficult. Before this, multitask learning was employed to analyze data from multiple sources simultaneously and acquire expertise in diverse domains. The EHR learning neural network model has trained multimodal neurons to perform general tasks, while some neurons are trained for a specific purpose. Activity can influence the results of laboratory tests and the types of data collected. Hidden discrete binary digits represent the Poisson distribution and parameters for each mode. The channel data is subsequently incorporated through the utilization of a feedforward network with shared hidden units [77].

8.9 CONCLUSION

The extensive and diverse clinical data stored in modern EHR systems may prove useful in identifying patients with varying health outcomes. Developing personalized phenotyping algorithms poses numerous obstacles for researchers, despite the plethora of information at their disposal. Machine learning techniques could potentially be valuable in addressing these challenges. To increase the adoption of machine learning for phenotyping tasks using EHRs, further efforts are required to develop machine learning algorithms that are adaptable and applicable in diverse environments. To optimize the utilization of machine learning algorithms in electronic phenotyping procedures, it is crucial to address problems regarding transparency and the size of training data, as these factors are essential for attaining favorable results.

KEYWORDS

- **convolutional neural network**
- **electronic health record**
- **electronic phenotyping**
- **health outcomes**
- **heart failure**
- **pattern mining techniques**

REFERENCES

1. McGinnis, J. M., Olsen, L., Goolsby, W. A., & Grossmann, C. (2011). *Clinical Data as the Basic Staple of Health Learning: Creating and Protecting a Public Good: Workshop Summary*. National Academies Press.
2. McCord, K. A., & Hemkens, L. G. (2019). Using electronic health records for clinical trials: Where do we stand and where can we go? *CMAJ*, *191*(5), E128–E133.
3. Davies, J., Sheridan, H., Bell, N., Cunningham, S., Davis, S. D., Elborn, J. S., Milla, C. E., Starner, T. D., Weiner, D. J., Lee, P. S., & Ratjen, F. (2013). Assessment of clinical response to ivacaftor with lung clearance index in cystic fibrosis patients with a G551D-CFTR mutation and preserved spirometry: A randomized controlled trial. *The Lancet Respiratory Medicine*, *1*(8), 630–638.
4. Beesley, L. J., Salvatore, M., Fritsche, L. G., Pandit, A., Rao, A., Brummett, C., Willer, C. J., Lisabeth, L. D., & Mukherjee, B. (2020). The emerging landscape of health research based on biobanks linked to electronic health records: Existing resources, statistical challenges, and potential opportunities. *Statistics in Medicine*, *39*(6), 773–800.

5. Liu, R., Rizzo, S., Whipple, S., Pal, N., Pineda, A. L., Lu, M., Arnieri, B., Lu, Y., Capra, W., Copping, R., & Zou, J. (2021). Evaluating eligibility criteria of oncology trials using real-world data and AI. *Nature*, *592*(7855), 629–633.
6. Geva, A., Abman, S. H., Manzi, S. F., Ivy, D. D., Mullen, M. P., Griffin, J., Lin, C., Savova, G. K., & Mandl, K. D. (2020). Adverse drug event rates in pediatric pulmonary hypertension: A comparison of real-world data sources. *Journal of the American Medical Informatics Association*, *27*(2), 294–300.
7. Rogers, J. R., Lee, J., Zhou, Z., Cheung, Y. K., Hripcsak, G., & Weng, C. (2021). Contemporary use of real-world data for clinical trial conduct in the United States: A scoping review. *Journal of the American Medical Informatics Association*, *28*(1), 144–154.
8. Boland, M. R., Hripcsak, G., Shen, Y., Chung, W. K., & Weng, C. (2013). Defining a comprehensive verotype using electronic health records for personalized medicine. *Journal of the American Medical Informatics Association*, *20*(e2), e232–e238.
9. Liao, K. P., Cai, T., Savova, G. K., Murphy, S. N., Karlson, E. W., Ananthakrishnan, A. N., Gainer, V. S., Shaw, S. Y., Xia, Z., Szolovits, P., & Churchill, S. (2015). Development of phenotype algorithms using electronic medical records and incorporating natural language processing. *BMJ*, *350*, h1885.
10. Samoutis, G., Soteriades, E. S., Kounalakis, D. K., Zachariadou, T., Philalithis, A., & Lionis, C. (2007). Implementation of an electronic medical record system in previously computer-naïve primary care centres: A pilot study from Cyprus. *Informatics in Primary Care*, *15*(4).
11. Pendergrass, S. A., & Crawford, D. C. (2019). Using electronic health records to generate phenotypes for research. *Current Protocols in Human Genetics*, *100*(1), e80.
12. Wang, F., Zhang, P., Qian, B., Wang, X., & Davidson, I. (2014). Clinical risk prediction with multilinear sparse logistic regression. In *Proceedings of the 20th ACM SIGKDD International Conference on Knowledge Discovery and Data Mining* (pp. 145–154).
13. Zhang, P., Wang, F., Hu, J., & Sorrentino, R. (2014). Towards personalized medicine: Leveraging patient similarity and drug similarity analytics. *AMIA Summits on Translational Science Proceedings*, *132*.
14. Wang, F., Sun, J., & Ebadollahi, S. (2012). Composite distance metric integration by leveraging multiple experts' inputs and its application in patient similarity assessment. *Statistical Analysis and Data Mining: The ASA Data Science Journal*, *5*(1), 54–69.
15. Tao, C., Parker, C. G., Oniki, T. A., Pathak, J., Huff, S. M., & Chute, C. G. (2011). An OWL meta-ontology for representing the clinical element model. In *AMIA Annual Symposium Proceedings 2011* (Vol. 2011, p. 1372).
16. Pathak, J., Kho, A. N., & Denny, J. C. (2013). Electronic health records-driven phenotyping: Challenges, recent advances, and perspectives. *Journal of the American Medical Informatics Association*, *20*(e2), e206–e211.
17. Zhou, J., Wang, F., Hu, J., & Ye, J. (2014). From micro to macro: Data-driven phenotyping by densification of longitudinal electronic medical records. In *Proceedings of the 20th ACM SIGKDD International Conference on Knowledge Discovery and Data Mining* (pp. 135–144).
18. Keravnou, E. T. (1996). Temporal diagnostic reasoning based on time-objects. *Artificial Intelligence in Medicine*, *8*(3), 235–265.

19. Wang, F., Lee, N., Hu, J., Sun, J., & Ebadollahi, S. (2012). Towards heterogeneous temporal clinical event pattern discovery: A convolutional approach. In *Proceedings of the 18th ACM SIGKDD International Conference on Knowledge Discovery and Data Mining* (pp. 453–461).
20. Miriovsky, B. J., Shulman, L. N., & Abernethy, A. P. (2012). Importance of health information technology, electronic health records, and continuously aggregating data to comparative effectiveness research and learning health care. *Journal of Clinical Oncology*, *30*(34), 4243–4248.
21. Amarasingham, R., Moore, B. J., Tabak, Y. P., Drazner, M. H., Clark, C. A., Zhang, S., Reed, W. G., Swanson, T. S., Ma, Y., & Halm, E. A. (2010). An automated model to identify heart failure patients at risk for 30-day readmission or death using electronic medical record data. *Medical Care*, *48*(11), 981–988.
22. Wang, X., Wang, F., Hu, J., & Sorrentino, R. (2014). Exploring joint disease risk prediction. In *AMIA Annual Symposium Proceedings 2014* (Vol. 2014, p. 1180).
23. Gotz, D., Wang, F., & Perer, A. (2014). A methodology for interactive mining and visual analysis of clinical event patterns using electronic health record data. *Journal of Biomedical Informatics*, *48*, 148–159.
24. Perer, A., & Wang, F. (2014). Frequence: Interactive mining and visualization of temporal frequent event sequences. In *Proceedings of the 19th International Conference on Intelligent User Interfaces* (pp. 153–162).
25. Ho, J. C., Ghosh, J., Steinhubl, S. R., Stewart, W. F., Denny, J. C., Malin, B. A., & Sun, J. (2014). Limestone: High-throughput candidate phenotype generation via tensor factorization. *Journal of Biomedical Informatics*, *52*, 199–211.
26. Zhou, X., Menche, J., Barabási, A. L., & Sharma, A. (2014). Human symptoms–disease network. *Nature Communications*, *5*, 4212.
27. Murdoch, T. B., & Detsky, A. S. (2013). The inevitable application of big data to health care. *JAMA*, *309*(13), 1351–1352.
28. Rait, G., Walters, K., Griffin, M., Buszewicz, M., Petersen, I., & Nazareth, I. (2009). Recent trends in the incidence of recorded depression in primary care. *The British Journal of Psychiatry*, *195*(6), 520–524.
29. Salama, R. A., Youssef, A., & Fahmy, A. (2018). Morphological word embedding for Arabic. *Procedia Computer Science*, *142*, 83–93.
30. Beaulieu-Jones, B. K., & Greene, C. S. (2016). Semi-supervised learning of the electronic health record for phenotype stratification. *Journal of Biomedical Informatics*, *64*, 168–178.
31. Baytas, I. M., Xiao, C., Zhang, X., Wang, F., Jain, A. K., & Zhou, J. (2017). Patient subtyping via time-aware LSTM networks. In *Proceedings of the 23rd ACM SIGKDD International Conference on Knowledge Discovery and Data Mining* (pp. 65–74).
32. Cheng, Y., Wang, F., Zhang, P., & Hu, J. (2016). Risk prediction with electronic health records: A deep learning approach. In *Proceedings of the 2016 SIAM International Conference on Data Mining* (pp. 432–440). Society for Industrial and Applied Mathematics.
33. Johnson, A. E., Pollard, T. J., Shen, L., Lehman, L. W., Feng, M., Ghassemi, M., Moody, B., Szolovits, P., Anthony Celi, L., & Mark, R. G. (2016). MIMIC-III, a freely accessible critical care database. *Scientific Data*, *3*(1), 1–9.

34. Kam, H. J., & Kim, H. Y. (2017). Learning representations for the early detection of sepsis with deep neural networks. *Computers in Biology and Medicine*, *89*, 248–255.
35. Che, C., Xiao, C., Liang, J., Jin, B., Zhou, J., & Wang, F. (2017). An RNN architecture with dynamic temporal matching for personalized predictions of Parkinson's disease. In *Proceedings of the 2017 SIAM International Conference on Data Mining* (pp. 198–206). Society for Industrial and Applied Mathematics.
36. Liu, B., Liu, J., Wang, G., Huang, K., Li, F., Zheng, Y., Luo, Y., & Zhou, F. (2015). A novel electrocardiogram parameterization algorithm and its application in myocardial infarction detection. *Computers in Biology and Medicine*, *61*, 178–184.
37. Vani, A., Jernite, Y., & Sontag, D. (2017). Grounded recurrent neural networks. *arXiv preprint arXiv:1705.08557*.
38. Mullenbach, J., Wiegreffe, S., Duke, J., Sun, J., & Eisenstein, J. (2018). Explainable prediction of medical codes from clinical text. *arXiv preprint arXiv:1802.05695*.
39. Abidi, M. H., Mohammed, M. K., & Alkhalefah, H. (2022). Predictive maintenance planning for industry 4.0 using machine learning for sustainable manufacturing. *Sustainability*, *14*(6), 3387.
40. Baumel, T., Nassour-Kassis, J., Cohen, R., Elhadad, M., & Elhadad, N. (2017). Multi-label classification of patient notes: A case study on ICD code assignment. *arXiv preprint arXiv:1709.09587*.
41. Yoon, H. J., Ramanathan, A., & Tourassi, G. (2017). Multi-task deep neural networks for automated extraction of primary site and laterality information from cancer pathology reports. In *Advances in Big Data: Proceedings of the 2nd INNS Conference on Big Data, October 23–25, 2016, Thessaloniki, Greece* (pp. 195–204). Springer International Publishing.
42. Percy, C., Holten, V. V., Muir, C. S., & World Health Organization. (1990). *International Classification of Diseases for Oncology*. World Health Organization.
43. Choi, E., Bahadori, M. T., Searles, E., Coffey, C., Thompson, M., Bost, J., Tejedor-Sojo, J., & Sun, J. (2016). Multi-layer representation learning for medical concepts. In *Proceedings of the 22nd ACM SIGKDD International Conference on Knowledge Discovery and Data Mining* (pp. 1495–1504).
44. Miotto, R., Li, L., Kidd, B. A., & Dudley, J. T. (2016). Deep patient: An unsupervised representation to predict the future of patients from the electronic health records. *Scientific Reports*, *6*(1), 1–10.
45. Gehrmann, S., Dernoncourt, F., Li, Y., Carlson, E. T., Wu, J. T., Welt, J., Foote Jr., J., Moseley, E. T., Grant, D. W., Tyler, P. D., & Celi, L. A. (2017). Comparing rule-based and deep learning models for patient phenotyping. *arXiv preprint arXiv:1703.08705*.
46. Zong, C., & Strube, M. (2015). *Proceedings of the 53rd Annual Meeting of the Association for Computational Linguistics and the 7th International Joint Conference on Natural Language Processing (Volume 1: Long Papers)*. In *Proceedings of the 53rd Annual Meeting of the Association for Computational Linguistics and the 7th International Joint Conference on Natural Language Processing (Volume 1: Long Papers)*.
47. Choi, E., Schuetz, A., Stewart, W. F., & Sun, J. (2017). Using recurrent neural network models for early detection of heart failure onset. *Journal of the American Medical Informatics Association*, *24*(2), 361–370.

48. Futoma, J., Morris, J., & Lucas, J. (2015). A comparison of models for predicting early hospital readmissions. *Journal of Biomedical Informatics, 56*, 229–238.
49. Rajkomar, A., Oren, E., Chen, K., Dai, A. M., Hajaj, N., Liu, P. J., Liu, X., Sun, M., Sundberg, P., Yee, H., & Zhang, K. (2018). Scalable and accurate deep learning with electronic health records. *NPJ Digital Medicine, 1*, 18.
50. Dernoncourt, F., Lee, J. Y., Uzuner, O., & Szolovits, P. (2017). De-identification of patient notes with recurrent neural networks. *Journal of the American Medical Informatics Association, 24*(3), 596–606.
51. Liu, Z., Tang, B., Wang, X., & Chen, Q. (2017). De-identification of clinical notes via recurrent neural network and conditional random field. *Journal of Biomedical Informatics, 75*, S34–S42.
52. Tompson, J. J., Jain, A., LeCun, Y., & Bregler, C. (2014). Joint training of a convolutional network and a graphical model for human pose estimation. *Advances in Neural Information Processing Systems, 27.*
53. Sutskever, I., Vinyals, O., & Le, Q. V. (2014). Sequence to sequence learning with neural networks. *Advances in Neural Information Processing Systems, 27.*
54. Hinton, G., Deng, L., Yu, D., Dahl, G. E., Mohamed, A. R., Jaitly, N., Senior, A., Vanhoucke, V., Nguyen, P., Sainath, T. N., & Kingsbury, B. (2012). Deep neural networks for acoustic modeling in speech recognition: The shared views of four research groups. *IEEE Signal Processing Magazine, 29*(6), 82–97.
55. Zhu, Z., Yin, C., Qian, B., Cheng, Y., Wei, J., & Wang, F. (2016). Measuring patient similarities via a deep architecture with medical concept embedding. In *2016 IEEE 16th International Conference on Data Mining (ICDM)* (pp. 749–758). IEEE.
56. Choi, Y., Chiu, C. Y., & Sontag, D. (2016). Learning low-dimensional representations of medical concepts. *AMIA Summits on Translational Science Proceedings, 2016*, 41.
57. Tierney, W. M., Takesue, B. Y., Vargo, D. L., & Zhou, X. H. (1996). Using electronic medical records to predict mortality in primary care patients with heart disease: Prognostic power and pathophysiologic implications. *Journal of General Internal Medicine, 11*, 83–91.
58. Suo, Q., Xue, H., Gao, J., & Zhang, A. (2016). Risk factor analysis based on deep learning models. In *Proceedings of the 7th ACM International Conference on Bioinformatics, Computational Biology, and Health Informatics* (pp. 394–403).
59. Yuan, Y., Xun, G., Jia, K., & Zhang, A. (2017). A multi-view deep learning method for epileptic seizure detection using short-time Fourier transform. In *Proceedings of the 8th ACM International Conference on Bioinformatics, Computational Biology, and Health Informatics* (pp. 213–222).
60. Prince, M., Patel, V., Saxena, S., Maj, M., Maselko, J., Phillips, M. R., & Rahman, A. (2007). No health without mental health. *The Lancet, 370*(9590), 859–877.
61. Lasko, T. A., Denny, J. C., & Levy, M. A. (2013). Computational phenotype discovery using unsupervised feature learning over noisy, sparse, and irregular clinical data. *PLOS ONE, 8*(6), e66341.
62. Lv, X., Guan, Y., Yang, J., & Wu, J. (2016). Clinical relation extraction with deep learning. *International Journal of Hybrid Information Technology, 9*(7), 237–248.
63. Lin, Q., Ye, S. Q., Huang, X. M., Li, S. Y., Zhang, M. Z., Xue, Y., & Chen, W. S. (2016). Classification of epileptic EEG signals with stacked sparse autoencoder based on deep learning. In *Intelligent Computing Methodologies: 12th International*

Conference, ICIC 2016, Lanzhou, China, August 2–5, 2016, Proceedings, Part III (pp. 802–810). Springer International Publishing.

64. Yan, B., Wang, Y., Li, Y., Gong, Y., Guan, L., & Yu, S. (2016). An EEG signal classification method based on sparse auto-encoders and support vector machine. In *2016 IEEE/CIC International Conference on Communications in China (ICCC)* (pp. 1–6). IEEE.
65. Veličković, P., Karazija, L., Lane, N. D., Bhattacharya, S., Liberis, E., Liò, P., Chieh, A., Bellahsen, O., & Vegreville, M. (2018). Cross-modal recurrent models for weight objective prediction from multimodal time-series data. In *Proceedings of the 12th EAI International Conference on Pervasive Computing Technologies for Healthcare* (pp. 178–186).
66. Bailey, J., Khan, L., Washio, T., Dobbie, G., Huang, J. Z., Wang, R., (Eds.). (2016). *Advances in Knowledge Discovery and Data Mining: 20th Pacific-Asia Conference, PAKDD 2016, Auckland, New Zealand, April 19–22, 2016, Proceedings, Part II.* Springer.
67. Pham, T., Tran, T., Phung, D., & Venkatesh, S. (2017). Predicting healthcare trajectories from medical records: A deep learning approach. *Journal of Biomedical Informatics, 69*, 218–229.
68. Choi, E., Bahadori, M. T., Sun, J., Kulas, J., Schuetz, A., & Stewart, W. (2016). Retain: An interpretable predictive model for healthcare using reverse time attention mechanism. *Advances in Neural Information Processing Systems, 29.*
69. Bahdanau, D., Cho, K., & Bengio, Y. (2014). Neural machine translation by jointly learning to align and translate. *arXiv preprint arXiv:1409.0473.*
70. Ayyar, S., Don, O., & Iv, W. (2016). Tagging patient notes with ICD-9 codes. In *Proceedings of the 29th Conference on Neural Information Processing Systems* (pp. 1–8).
71. Lipton, Z. C., Kale, D. C., Elkan, C., & Wetzel, R. (2015). Learning to diagnose with LSTM recurrent neural networks. *arXiv preprint arXiv:1511.03677.*
72. Ma, F., Chitta, R., Zhou, J., You, Q., Sun, T., & Gao, J. (2017). Dipole: Diagnosis prediction in healthcare via attention-based bidirectional recurrent neural networks. In *Proceedings of the 23rd ACM SIGKDD International Conference on Knowledge Discovery and Data Mining* (pp. 1903–1911).
73. Goodwin, T. R., & Harabagiu, S. M. (2017). Deep learning from EEG reports for inferring underspecified information. *AMIA Summits on Translational Science Proceedings, 2017*, 112.
74. Suresh, H., Szolovits, P., & Ghassemi, M. (2017). The use of autoencoders for discovering patient phenotypes. *arXiv preprint arXiv:1703.07004.*
75. Alaa, A. M., Weisz, M., & Van Der Schaar, M. (2017). Deep counterfactual networks with propensity-dropout. *arXiv preprint arXiv:1706.05966.*
76. Henao, R., Lu, J. T., Lucas, J. E., Ferranti, J., & Carin, L. (2016). Electronic health record analysis via deep Poisson factor models. *The Journal of Machine Learning Research, 17*(1), 6422–6453.
77. Torres, J. F., Hadjout, D., Sebaa, A., Martínez-Álvarez, F., & Troncoso, A. (2021). Deep learning for time series forecasting: A survey. *Big Data, 9*(1), 3–21.

Index

L

M

Q

R

S

T

U

V

W

X

Y

Z